25 at 25:
A selection of articles from the SCSC Newsletter Safet

Related Titles

Improvements in System Safety
Proceedings of the Sixteenth Safety-critical Systems Symposium, Bristol, UK, 2008
Redmill and Anderson (Eds)
978-1-84800-099-5

Safety-Critical Systems: Problems, Process and Practice
Proceedings of the Seventeenth Safety-critical Systems Symposium, Brighton, UK, 2009
Dale and Anderson (Eds)
978-1-84882-348-8

Making Systems Safer
Proceedings of the Eighteenth Safety-critical Systems Symposium, Bristol, UK, 2010
Dale and Anderson (Eds)
978-1-84996-085-4

Advances in Systems Safety
Proceedings of the Nineteenth Safety-critical Systems Symposium, Southampton, UK, 2011
Dale and Anderson (Eds)
978-0-85729-132-5

Achieving Systems Safety
Proceedings of the Twentieth Safety-critical Systems Symposium, Bristol, UK, 2012
Dale and Anderson (Eds)
978-1-4471-2493-1

Assuring the Safety of Systems
Proceedings of the Twenty-first Safety-critical Systems Symposium, Bristol, UK, 2013
Dale and Anderson (Eds)
978-1481018647

Addressing Systems Safety Challenges
Proceedings of the Twenty-second Safety-critical Systems Symposium, Brighton, UK, 2014
Dale and Anderson (Eds)
978-1491263648

Engineering Systems for Safety
Proceedings of the Twenty-third Safety-critical Systems Symposium, Bristol, UK, 2015
Parsons and Anderson (Eds)
978-1505689082

Developing Safe Systems
Proceedings of the Twenty-fourth Safety-critical Systems Symposium, Brighton, UK, 2016
Parsons and Anderson (Eds)
978-1519420077

Developments in System Safety Engineering
Proceedings of the Twenty-fifth Safety-critical Systems Symposium, Bristol, UK, 2017
Parsons and Kelly (Eds)
978-1540796288

Mike Parsons • Graham Jolliffe • Tim Kelly
Editors

25 at 25:

A selection of articles from twenty-five years of the SCSC Newsletter Safety Systems

Safety-Critical
Systems Club

SCSC-136

The publication of these articles is sponsored by BAE Systems plc and Jaguar Land Rover

Editors

Graham Jolliffe	Mike Parsons	Tim Kelly
13 Irving Road	NATS CTC	Department of Computer Science
Southbourne	4000 Parkway	University of York
Bournemouth	Whiteley, Fareham	Deramore Lane, York
Dorset, BH6 5BG	PO15 7FL	YO10 5GH
United Kingdom	United Kingdom	United Kingdom

ISBN-13: 978-1540896483
ISBN-10: 154089648X

© Safety-Critical Systems Club 2017. All Rights Reserved

Individual articles © as shown on respective first pages

Preface

This book contains selected articles from 25 years of the SCSC Newsletter, Safety Systems for this year's Silver Jubilee of the Safety Critical Systems Club.

Safety Systems was edited by Felix Redmill from its start in 1991 until mid-2016. Felix produced an informative and thought-provoking publication with many important contributions. Felix managed to commission high-quality articles from both established and new authors over the years, and deliver these in a format which became a valuable resource.

The newsletter tracked many events and influences over the years, for instance the Year 2000 issue and the development of ISO 26262.

The articles presented here were selected for the Club's 25th anniversary to show the breadth and depth of the articles included in the newsletter. They were chosen personally by the editors, based on historical or topical influence, because they were later shown to be important, or simply because they were well-written or amusing. This selection is in no way meant to be complete or comprehensive.

Katrina Attwood at the University of York has now taken on the editorship of Safety Systems and is keen to continue Felix's good work. She is always looking out for articles. Please contact her at katrina.attwood@scsc.uk if you are interested in submitting a contribution.

We are grateful to our sponsors BAE Systems and Jaguar Land Rover for their valuable support of the club.

MP, GJ & TK

Please note that authors' affiliations and biographies have been given as they appeared in the original articles; they may well have changed in the intervening years. Contact details for authors have been omitted as many included telephone numbers.

Copyright of the articles is retained by the original copyright owners; in most cases this is the author.

A message from the sponsors

BAE Systems and Jaguar Land Rover are pleased to support the publication of these articles. We recognise the benefit of the Safety-Critical Systems Club in promoting safety engineering in the UK and value the opportunities provided for continued professional development and the recognition and sharing of good practice. The safety of our employees, those using our products and the general public is critical to our business and is recognised as an important social responsibility.

The Safety-Critical Systems Club

publisher of the

Safety Systems Newsletter

Safety-critical systems and the accidents that don't happen

When an aircraft crashes, it makes headlines. That hundreds of thousands of flights each week do not crash is accepted as routine. Airliners, air traffic control systems, railway signalling, car braking systems, defence systems, nuclear power stations and medical equipment are some of the critical systems in use, on which life and property depend. New autonomous systems that will affect our daily life are coming on-stream soon, including delivery drones and self-driving road vehicles. That safety-critical systems do work well is because of the expertise and diligence of professional systems safety engineers, regulators and other practitioners who work to minimise both the likelihood that accidents will occur, and the consequences of those that do. Their efforts prevent untold deaths and injuries every year. The Safety-Critical Systems Club (SCSC) has been actively engaged for over twenty-five years to help to ensure that this continues to be the case.

What is the Safety-Critical Systems Club?

The SCSC is the UK's professional network and community for sharing knowledge about safety-critical systems. It brings together engineers and specialists from a range of disciplines working on safety-critical systems in a wide variety of industries, academics researching in the field, providers of the tools and services that help develop the systems, and the regulators who oversee safety. It provides, through publications, seminars, tutorials, a website, working groups and, importantly, at the annual Safety-critical Systems Symposium, opportunities for them to network and benefit from each other's experience in working hard at the accidents that don't happen. It focuses on current and emerging practices in safety engineering, software engineering, and product and process safety standards.

What does the SCSC do?

The SCSC maintains a website (www.scsc.uk), which includes directories of tools and services that assist in the development of safety-critical systems. It publishes a regular newsletter, Safety Systems, three times a year from which these articles are selected. It organises seminars, workshops and training on general matters or specific subjects of current concern, which are prepared and led by world experts. Since 1993 it has organised the annual Safety-critical Systems Symposium (SSS) where leaders in different aspects of safety, from different industries, including consultants, regulators and academics, meet to exchange information and experience, with the papers published in a proceedings volume. The SCSC supports industry working groups, such as the Data Safety Initiative Working Group (DSIWG) that is addressing concerns raised about data in safety-related systems. New working groups on Autonomy and Service Assurance are starting in 2017 and others are planned. The SCSC carries out all these activities to support its mission:

> ... to raise awareness and facilitate technology transfer in the field of safety-critical systems ...

History

The SCSC began its work in 1991, supported by the Department of Trade and Industry and the Engineering and Physical Sciences Research Council. The Club has been self-sufficient since 1994, but enjoys the active support of the Health and Safety Executive, the Institution of Engineering and Technology, and the BCS - The Chartered Institute for IT; all are represented on the SCSC Steering Group.

Membership

Membership may be either corporate or individual. Individual membership entitles the member to Safety Systems three times a year, other mailings, and discounted entry to seminars, workshops and the annual Symposium. Frequently individual membership is paid by the employer.

Corporate membership is for organisations that would like several employees to take advantage of the benefits of SCSC programmes. The amount charged is tailored to the needs of the organisation.

Individual membership can be obtained online at: http://www.scsc.uk/

For more information about membership, please contact: Alex King, Department of Computer Science, University of York, phone: 01904 325402 email: alex.king@scsc.uk or mike.parsons@scsc.uk

Contents

Is this a ~~dagger~~ Jumbo which I see before me?
Ken Frith 1

Software safety - by prescription or argument?
Tim Kelly 5

Risking Management
John Ridgway 9

People 5, Techniques 2, Tools 1 and no Penalties?
Stan Price 15

The Need for New Paradigms in Safety Engineering
Nancy G. Leveson 19

Medical radiological incidents: the human element in complex systems
Chris Moore 25

The ALARP Principle: Legal and Historical Background
Felix Redmill 31
Making ALARP Decisions
Felix Redmill 39
Efficiency and Effectiveness of the ALARP Principle
Felix Redmill 53

Armageddon and Other Failure Modes
John Ridgway 59

Software testing and IEC 61508 – a case study
Ian Gilchrist 65

IEC 61508 and Related Guidance - Uses and Abuses
David J Smith 73

Human factors — how little can you get away with, and how much is right?
Brian Sherwood-Jones 79

Application of Formal Methods in a Commercial Environment
Guy Mason 83

A discussion of risk tolerance principles
Odd Nordland 89

Quality assurance of software used in diagnostic medical imaging
Philip Cosgriff 95

There's No Substitute for Thinking
Tim Kelly 99

The new MISRA documents
David Ward 105

ISO DIS 26262 The new automotive functional safety standard
David D Ward 111

Review: IRSE Guidance on the Application of Safety Assurance Processes in the Signalling Industry (May 2010)
Andrew Rae and Mark Nicholson 117

Complexologistification
Rob Collins 127

Reply to Complexologistification
Kevin Geary 131

Demonstrating safety in global navigation systems
Steve Leighton 133

Software as goods: some answers, but yet more questions
Ian Lloyd 137

Safety and the Year 2000
Tony Foord 141

The millennium timebomb (the Y2K problem) — a consultant's dream, or a real problem?
Ray Ward 147

Reverse Engineering the Software Design of a Safety-related ATM System
Ron Pierce and Mary Johnston 151

Author Index 157

Is this a ~~dagger~~ Jumbo which I see before me?

Ken Frith

Crew Services Ltd

First published in Safety Systems 5-1, September 1995.

I have a problem to set before you. Regrettably it is not the sort of problem where you can turn to the staff answer at the foot of the page when you get stuck, but is one which, I believe, has no clear solution at present. I am thus hoping that some of the experts reading this newsletter may cast some fresh thought on the matter.

This problem involves a cliff-hanger worthy of an episode of a certain well-known space odyssey. Picture this: The starship Captain is poised in front of his viewing screen on which he sees an unidentified object hurtling directly towards him. His sensors offer little information except for the dynamics of the object, and at his side his trusty lieutenant is saying to him, 'Well Captain, there is an 89.8% chance that it is hostile, a 5.7% chance that it is another Starfleet cruiser, and a 4.5% chance that it is a neutral passenger shuttle.

Return to the present. You are no longer in a fantasy but in a real situation: you are in the operations room of a naval warship off the coast of Bosnia. You are not at war, but giving peace-time support to the United Nations. Intelligence sources tell you that at least one of the parties in the conflict possesses anti-ship missiles, and has the political will or instability to use them. The area is congested with friendly units ferrying stores and evacuees, and is overflown by friendly aircraft and commercial traffic, plus (inevitably these days), chartered aircraft filled with the world's press; life must go on. You too have an unidentified contact on your screen, approaching your ship at an alarming rate and in a distinctly disturbing manner. Now the safety element - you have to make a decision, under extreme pressure and in a very short space of time that will inevitably put lives at risk. Do you take action, with a distinct possibility that you may destroy a friendly unit — or worse, a civilian airliner? Or do you delay, with a comparable probability that the object is an incoming missile?

This may seem sensationalist, but there is a serious point. How are the safety issues of complex man-machine systems to be addressed? To naval commanders this is not a hypothetical question, but one which they regularly have to face. For a few, their decisions have been wrong. I cite two examples: in the first, a US cruiser shot down an Iranian Airbus with the loss of over two hundred civilian lives, in the mistaken belief that it was an attack on itself by hostile Iranian forces; in the second, a US destroyer was hit by two Exocet missiles from an Iraqi (then friendly) warplane, having failed to take timely action in the mistaken belief that the incoming object was not a threat. Although the first was an error of commission, and the second one of omission, both of these incidents resulted from incorrect decisions by the ship's commander, both involved multiple loss of life (i.e., accident severity 'catastrophic'), and both had a high probability of occurrence (probability category 'frequent' or 'probable'). Hence we have systems with a risk class of 'Intolerable' that are clear candidates for extensive risk mitigation, or which could never be given safety clearance for user.

In the past, such decision making has been the responsibility of the human element in the system - usually the commander. Also in the past, this problem has not been treated as a specific safety issue. However, we are now entering an era where system designers are required to address the system (including the human element) as a whole, and also to consider every eventuality that might lead to an accident.

Firstly, let us look at the systems. A naval warship is a complex system made up of a number of constituent complex sub-systems and is itself part of a wider complex system (the force of which it is part). The warship differs from most other safety-related systems in that, far from being designed not to cause harm, it is specifically designed to do so. The safety element is to ensure that it harms only those objects designated by the user. But here is the catch: the user is of course a part of the system!

The combat systems of a warship consist typically of sensor sub-systems (radars, electrooptics, electronic and communications intercept systems and data links from other platforms), effector sub-systems (guns, missiles and decoys) and Command, Control and Information (C2I) sub-systems. The first two groups involve relatively straightforward safety-critical systems, where safety analysis must concentrate on traditional hazards such as explosives, propellants, inflammables, remotely-controlled and automatic moving machinery, high-power radio and radar emitters, toxic substances and pressure systems. Although no safety-critical system can be regarded as routine, existing procedures in general ensure that these can individually be designed to an adequate level of safety

The problems arise in C2I sub-systems, which contain, inter alia, communications, information databases, decision aids and human computer interfaces (HCIs). Such systems may be centrally located or distributed around constituent sub-systems (which can be autonomous with their own C2 elements). C2I systems are different in that they are non-deterministic, they involve management rather than control functions, and they contain knowledge-based techniques and decision-making functionality.

Prior to the use of computers, much of this functionality resided exclusively in the human element of the system. However, the increased pace of modern warfare gives the modern commander mere seconds (rather than minutes or hours) to respond, and system complexity has become too great for the human to cope with unaided. Clearly the commander needs machine assistance to achieve the performance requirements, and it is often necessary for the system to bypass him in order that the delays resulting from human interaction do not preclude the achievement of satisfactory performance. In addition, we have unreliable source information - not necessarily due to poor intrinsic reliability of the sensors, but because the environment in which they operate is a world of uncertainty. The biggest problem is identification, usually achieved by a combination of separate methods - secondary radar, visual, electro-optic, intelligence sources and behavioural study. All of these can be variously unreliable, very often with an adversary intentionally making them so.

In the past, it has been the practice to place the onus on the human elements of the system to make value judgements of the situation, and then of course to blame them if they get it wrong. As illustrated in many circumstances, the human cannot alone now cope with the pace and complexity of this decision-making; nor is he likely to prove any more reliable in the future. The illuminating table in Issue 1 of Def Stan 00-56 (I regret that it has been removed from Issue 2) states that the probability of human operator unreliability (the failure to act correctly in reasonable time) can approach unity after the onset of a high-stress condition. Thus we have unreliable information feeding an unreliable decision process, controlling a highly reliable lethal system! How can we possibly make a safety case on that basis?

Technology and knowledge-based techniques now allow much more of the decision processes to be performed within the machine part of the system. However, formal demonstration of the safety case is now a mandatory requirement, and in this cosmopolitan world transgressions will not be tolerated, as they have been in the past. We therefore have a problem, as yet without solution, which could render modern weapons systems unusable by anybody except the out-and-out reckless, or in circumstances such as full-scale hostilities where safety cases could conceivably be ignored. You could dismiss this problem as one that merely demonstrates the futility of developing weapons of war, and you may have a point. But governments and society require us to attempt a resolution. In any case, the problems of presenting unreliable information to an unreliable human are not restricted to combat systems (I have in mind the average motorist on a foggy day). So I would welcome your thoughts on the subject - there is a very fine line between achieving a correct action and one that is disastrous

Software safety - by prescription or argument?

Tim Kelly

Rolls-Royce Systems and Software Engineering UTC
University of York

First published in Safety Systems 7-2, January 1998

At a recent workshop on industrial experience of safety-critical systems, held in Australia, the topic for debate in the panel discussion was 'Software Safety Standards'. The panel included industrial representation from aerospace (Praful Bhansali — Software Safety and Security Focal, Boeing), defence (Tony Cant —Defence Science and Technology Organisation) and the railways (Charles Page —Westinghouse Signals). The software safety research perspective was provided by Jim Welsh (from the Software Verification Research Centre, University of Queensland) and myself (representing Rolls–Royce plc and the High Integrity Systems Engineering Group at York). The discussion particularly focussed on the following questions:

- how do software safety standards differ from traditional safety standards?
- what is the role of the software safety case?

Addressing the first of these questions, the discussion highlighted the fact that there has been an intentional shift in the role of traditional safety standards, away from prescription (telling the developer how to build their system safely) to target setting (forcing the developer to argue how they have achieved the appropriate targets). (This is certainly true of, for example, the U.K. Health and Safety Executive's approach to certification of the off-shore and railways industries.) However, with software safety standards (with both the U.K. Defence Standard 00-55 and IEC 1508 cited as examples) we appear to be prescribing to a great level of detail what programming languages can be safely used, the sufficient level of testing, etc. Nowhere is this more apparent than with the ubiqui-

© Tim Kelly 1998.
Published by the Safety-Critical Systems Club. All Rights Reserved

tous Software Integrity Levels. These provide a perfect example of where (we assume) the standards–writers appear to have 'done our thinking for us', e.g. concerning:

- how levels of integrity relate to target failure rates;
- how integrity levels can be apportioned through system decomposition;
- the appropriate development and assurance techniques for each integrity level;
- the level of claim we can make having followed the integrity level guidelines.

As no justification is provided of the requirements that emerge in the standards, it appears that the developer's role is simply to accept such things as givens. Although some may say that this is within the nature of standards (i.e. to state, not explain) it was the opinion of the discussion that it can be too easy to comply with the standards whilst missing their (undisclosed) intent. Also, it was highlighted that this approach can present prohibitively harsh requirements for small-budget projects where, perhaps, given the underlying rationale, requirements could justifiably be relaxed.

The answer? There appeared to be some consensus that where possible we should attempt to move away from prescription as our approach to ensuring software safety. Alternatively, more effort should be spent in determining the true, fundamental, requirements (targets) of software safety (for example, rather than specifying in detail the tools, techniques and languages of an 'Integrity Level 4 process', instead working out the required attributes of a process to meet Integrity Level 4). The responsibility should then be placed on the software and system developers to present an argument as to why their systems are safe — how they meet the fundamental intent of the standards. It was suggested that such an approach could provide benefit with respect to safety — by making it harder to comply with standards in an 'unthinking' manner — and financially — by permitting arguments of appropriateness, and reduction of risks 'as low as reasonably practicable' (ALARP).

It would wrong to pretend that amongst all this 'idealism' that there wasn't an underlying pragmatic concern from those representing industry. Specifically, they wanted standards to 'tell them what to do' so that they didn't have to spend time and money determining what standards meant only to be told that they had got it wrong! It was also recognised that even with an ALARP approach, limits must be set on what is considered tolerable. The conclusion was therefore that although the current level of prescription may be too great, a careful balance must be maintained between the prescription and argument/ target-setting approaches.

The role of the software safety case

Arising out of discussion of the first question, it was clear that we had begun to answer the second — concerning the role of the software safety case. As with the role of 'traditional' safety cases, the purpose of the software safety case is very definitely to present a convincing, comprehensive and defensible argument that the system (in this case, the software) is acceptably safe to operate within a given operational context.

Conversely, we appreciated that it is not the job of the software safety case simply to be the container for all the evidence required for compliance with the detailed requirements of the safety standards. Evidence of compliance alone does not a safety case make! In arguing the safety of our software components, we must 'get our hands dirty' with the details of the hazards of the systems in which they're placed. This means, first of all presenting an argument structured around sound and rigorous software hazard analyses (e.g. Software Hazards and Operability Studies as described by U.K Defence Standard 00-58). Secondly we must present, not in vague or ambiguous terms, but in a clear and specific manner, how the evidence available sufficiently addresses all of the software hazards identified. This reinforced Praful Bhansali's concern that we must be very clear how the effort spent in developing and testing software has actually contributed to assuring safety. His challenge to anyone suggesting that he use a new safety analysis technique was that they should clearly explain why he needed it, i.e. where did it fit into the overall safety argument?

The question of the defensibility of the safety case was also raised by Dr Tony Cant. His experience of safety cases suggested that a problematic area was the use of quantitative models, particularly:

- the belief that a problem is well understood if it is quantified;
- insufficient 'thought' and effort on those areas that had been quantified;
- unrealistic simplifications of a system in order to fit a model;
- inappropriate use of the figures contained within or resulting from a quantitative model.

The conclusion arising from the resulting discussion was that the use of probabilities in the software safety case must always be fully justified, and that although attractive (we all like numbers, don't we?) it was often an inferior substitute for well structured qualitative reasoning.

Some Conclusions

The overriding feeling surrounding the various issues discussed was that building safety-critical software is a uniquely difficult task. There are no easy answers — we should not look to standards wholly to prescribe an approach. We should be careful of over-simplification when translating qualitative problems (i.e. hazards) into the quantitative domain. Instead, our effort should be focussed on improving our capability to argue, lucidly and cogently, the safety case for our software systems.

Risking Management

John Ridgway

First published in Safety Systems 22-3, May 2013

The meeting was not going well. Everywhere I looked directors and managers were pointing their fingers at me and howling with uncontrolled laughter. To my right, the managing director shook helplessly and gripped his side as tears rolled down his pitted cheeks. To my left, the marketing director appeared to be on the point of soiling his satin trousers. I alone sat with a face of stone. The joke was on me.

What, you might ask, had placed me at the epicentre of such executive ridicule? Well, as the newly appointed quality manager, I had chosen the occasion of my first boardroom meeting to table an audit report documenting that the project under review had proceeded to the code-and-test phase without having received design approval. The audit report did not waste words: 'Nature of Non-Conformity: Absence of approved design. Proposed Corrective Action: Establish design approval.'

It mattered not that the non-conformity represented a potentially disastrous situation for a critical project. It mattered not that the remedy was already urgent and likely to be very expensive. What mattered that day, as far as my boardroom colleagues were concerned, was that the apparent banality of the report had confirmed all their suspicions regarding the calibre of intellect possessed by quality management professionals. Clearly, I was a fool amongst men, and this revelation had been the source of much merriment and mirth. After what seemed an eternity, the managing director recovered sufficient composure to put voice to the mocking of his managerial pack. 'Welcome John', he sniggered, 'to the fast-track of quality management'. And with that proclamation, he was able to regain control of the meeting and move it on to the serious question of whether the coding and testing were going to finish on time.

I have taken the time to recount this tale, since I think it provides a good example of one of the many respects in which managers can not only fail to manage the risks to their business, they can actually be the prime source of such risk.

If one accepts that the functioning of a management team can be compared to that of a system, then it follows that the team's dysfunction can be investigated using the concepts of systems analysis. Furthermore, insofar as the team has

© John Ridgway 2013
Published by the Safety-Critical Systems Club. All Rights Reserved

the responsibility to control the destiny of the organisation under their charge, it follows that failure of the management system can be analysed using the methods and techniques of risk analysis.

This idea is not new, of course. The editor of this very newsletter uses it as a central tenet when recommending that the safety risks posed by managers should be formally addressed whilst constructing safety cases (Redmill [2006]). By writing this article, I wish to fully endorse such views. You can see the evidence for the argument in a host of accident investigation reports, but for me the phenomenon has a more visceral and personal significance. Take it from me, a quality manager's job can get decidedly uncomfortable when project managers are being advised by their senior executive that 'those who cannot run rings around the quality manager are clearly not the men for the job'. You don't need a PhD in root-cause analysis to see where the problem lies here, although implementing a solution may prove easier said than done. The challenge is twofold: firstly, when attempting to investigate systemic issues, an analytical approach will be required and so choosing the correct analytical tools will be important. Secondly, some of the executive shortcomings that your analysis may reveal could be of the type that renders such executives incapable of understanding or accepting the analysis. Indeed, you may need a strategy for dealing with a hostile response.

It is to the problem of choosing a suitable analytical approach that I next turn, since I think it is the easier of the two challenges to deal with. I will then conclude the article with one or two personal views regarding the politics and practicalities of addressing management risk.

Choosing a Methodology

Business analysis is another of those fields that is thoroughly overburdened with methodologies, frameworks, modelling syntaxes, tools and, dare I say it, business analysts. Each methodology, etc., has an army of enthusiastic supporters who will swear by the power and utility of their personal favourite, and who am I to rain on their parade? I imagine that each method has something to commend it.

Many of the goodies on offer are quite grand and sophisticated [Note 1], but the maxim I would apply is 'keep it simple'. I have found that a lot can be captured regarding a business system simply by using IDEF0 (Integrated DEFinition methods) to illustrate business functions and their relationships [Note 2]. Each function (or activity, to use the IDEF jargon) is represented as shown in Figure 1.

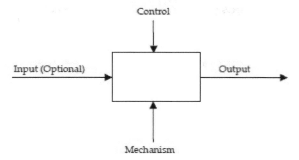

Fig.1. Process Representation (IDEFO)

You will note that this is very similar to the representation recommended by Redmill (see Figure 2). The main difference lies in the representation of functional relationships. The IDEF0 representation includes an arrow to represent control, and the fact that this control may be realised through the exchange of information is treated as circumstantial. In Figure 2, a similar arrow is used to represent exchange of information, and the fact that this may have a controlling effect is treated as consequential.

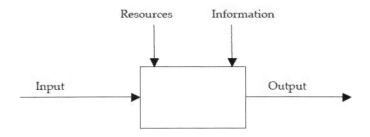

Fig.2. Process Representation (Redmill)

The distinction is largely academic. Nevertheless, my preference is to make the representation of control explicit, since failure to implement adequate or reliable controls is so often the real issue when it comes to business process failures (as it is when dealing with physical systems and their safety functions). Another reason for my penchant for IDEF0 is its explicit representation of each function's enabling mechanisms (for example, mechanisms related to work environment and resources). The provision of control and enablement pretty much sums up the purpose of management, so any representation that focuses on these two will help you identify the key areas of potential management failure. On a final note, IDEF0 is a commonly encountered method that is supported by a myriad of tools.

Adopting a Systematic Approach

IDEF0 is a top-down analytical method that can be used to diagrammatically represent business functions and their relationships. The whole business is first represented by a top-level diagram comprising a single activity (the so-called A0 activity) and then broken down into a hierarchical set of diagrams, each of which looks something like the Waterfall Model familiar to software developers. The main problem I have found with IDEF0 is that these diagrams can get pretty busy, pretty quickly. One is then left struggling over whether to have an accurate representation or one that can be readily understood. However, for our purposes here, we don't need to worry too much about diagrammatic representation. What we want is to be able to perform a FMEA (failure modes and effects analysis) on each featured activity, working methodically through all potential relationships with other activities. For this, all you need to do is list the activities in a spreadsheet and then ask the following question for each listed entry:

1. What are the inputs, controls and enabling mechanisms [Note 3] for the activity and how might the activity be adversely affected if they are inadequate or inappropriate?
2. What other activities are controlled by this activity, and what are the implications of this control proving inadequate or inappropriate?
3. In what way does this activity provide enabling mechanisms for other activities, and what are the implications of these mechanisms proving inadequate or inappropriate?
4. What other activities depend upon this activity for their inputs, and what are the implications of these inputs proving inadequate or inappropriate?

Of course, as you proceed, answers to questions 2-4 will start to inform your answers to question 1. The issue is that many of the activities concerned are undertaken by management, and the implications of inadequate outputs may not fully emerge until somewhat down the line, i.e. at the point of system or service delivery, or during system operations.

Here's One I Prepared Earlier

When it comes to deciding what functions [Note 4] to include in your process model, you don't have to start with a blank page. A number of public-domain frameworks, standards and process models have been developed for the purposes of organisational assessments and/or to assist with the development of management systems. Principal amongst these are TickIT, ITIL, PRINCE2, ISO/IEC 15504 and CMMI.

Most recently, the TickITplus scheme has been launched, and this incorporates a Base Process Library (BPL) amalgamating the essential features of ISO/IEC 12207, ISO/IEC 15288, ISO 20000-1 and ISO 27001. Many of the processes defined within the TickITplus BPL are particularly germane to the responsibilities and deeds of senior management, whilst also being clear candidates for a safety manager's attentions (the BPL processes referred to as 'Management of Infrastructure and Work Environment', 'Resource Management', 'De-cision Management', 'Risk Management', 'Corporate Management & Legal', and 'Business Relationships Man-agement' come to mind - and that's before we even start on the BPL's technical processes such as 'Configuration Management' and 'Stake-holders' Requirements Defin-ition'). In accordance with TickITplus, those BPL processes that are relevant to your organisation are used to construct your organisation's Process Reference Model (PRM). You can then use FMEA to analyse your PRM, paying particular attention to how each process's outputs serve to control and enable related business processes and how this might ultimately result in management failures.

However, please don't make the mistake of treating any of the above standards and frameworks with any great respect. By all means, cherry-pick them for ideas, but don't be afraid to add anything that you think adds grist to the analyst's mill or to modify them to better reflect the real world you inhabit. For example, if you think that promoting the right culture is an overarching concern (as indeed you should), then feel free to add this activity to the list. Correct me if I'm wrong, but I don't think this is itemised in any of the process lists provided by the 'off-the-shelf' process models.

But Don't Try This At Home

As my opening anecdote illustrated, boardroom culture can be the root cause of a whole raft of operational difficulties. However, the sad fact is that anyone within an organisation who is sufficiently distanced from the senior management systems under analysis as to be able to claim objective detachment is also likely to be in no position to comment with impunity. Furthermore, as I have already said, many of the executive traits and shortcomings that analysis might uncover are likely to be of the type that renders executives incapable of accepting such an analysis (for example, resistance to criticism from a β male may be such a trait). So where does that leave us?

Well it certainly doesn't leave us with the mandate to give up. The issue is far too important for that. I have shown above that there is a strong case for internal risk analysis, but if that is found not to be possible, we have to believe there are external options available. Third-party assessments are on offer, but the external organisations concerned are often compromised by commercial interests. Shareholders can exercise their influence through their demands for corporate governance but, in my experience, this is more effective in changing boardroom rhetoric than in changing the reality on the ground. Ultimately, the threat of legislation may be the only effective sanction, but this presupposes that executives can imagine themselves falling foul. I know a number of Health & Safety colleagues who say that tales of corporate manslaughter can be very effective in gaining an executive's attention. This will certainly focus their minds on the potential failings of employees, but that doesn't necessarily mean that they will start to see themselves as the prime liability.

I honestly don't know what the answer is. I just know that anyone who chooses to encompass senior management within their root cause analysis will need to share their sense of humour.

Footnotes

[1] Take, for example, Enterprise Architecture (EA) frameworks such as Zachman and The Open Group Architecture Framework (TOGAF 9). They are so expansive and convoluted that you can get qualifications in them. Nice!
[2] IDEF0 is used to represent functions and their relationships, but not workflow. For that you will need something like IDEF3. For a full description of the IDEF family of methods, see www.idef.com
[3] Note that the term 'input' is used by IDEF0 in a restricted sense. One should only use it to refer to those things that are consumed by the activity concerned, or things that do not survive in their initial form (i.e. the raw materials). Such input is commonplace within manufacturing industries but less so within industries that specialise in the generation and exploitation of intellectual capital. In the latter instance, you may find that many activities have controls and enabling mechanisms, and yet have no inputs!
[4] I am aware that I am using terms like 'function', 'activity' and 'process' interchangeably, and some of you may object to this. But please don't bother. I am retired now and so I no longer have to pretend to care about such things.

Reference

Redmill, Felix [2006] 'Understanding the Risks Posed by Management', In Redmill F and Anderson T (Eds.): Developments in Risk- Based Approaches to Safety: Proceedings of the Fourteenth Safety-critical Systems Sym-posium, Bristol, UK.

John Ridgway is a recently retired analyst whose 10 years' experience as a Quality Manager in a software and systems house left him with no doubt that senior management delinquency was the most significant risk factor affecting an organisation's ability to meet corporate governance obligations.

People 5, Techniques 2, Tools 1 and no Penalties

Stan Price

Price Project Services Ltd

First published in Safety Systems 6-1 September 1996.

Many years ago, when England won soccer World Cups, I was a programmer. The user, an experimental aircraft project engineer working on bizarre designs such as blown pole rotors, used to bring me his requirements literally on the back of a fag packet, Capstans I think. I would code them and then, after lengthy periods of debugging, the user would look at the results and change the requirements.

Experience brought with it the confidence to challenge this way of working and I was soon insisting he thought out his requirements in advance and present them in a typed document following a format I dictated. The user, up to then, had been a friend.

The loss of this friend did not matter so much as I was soon in charge of legions of programmers all of whom had to follow the nice tidy procedures I dreamt up to make my world much more tidy, hence visible, hence controllable. Exposure to other people's ideas, plus my own innovation, made these procedures more and more complex. I felt quite comfortable about this as others were doing the same thing and I was becoming immersed in a world with its own supporting jargon, waterfall model, top down design, etc., etc.

I even remember being green with envy of a large West Coast electronics company when I met one of its employees who announced that he was head of the section that monitored test coverage metrics for Pascal code!

However, late in life, over the last decade I have begun to doubt this complexity. This has been brought about by the remarkable observation that project success is not brought about by this complexity and, indeed, that projects succeed or fail quite independently of it. The only factors that ensured success were realistic goals with happy competent teams dedicated to meeting them.

I have therefore for some years been working on a methodology based primarily on human and social factors. I call it a methodology to show that I am not totally oblivious of current fashion.

Surprisingly, this has its critics. The most prominent statements that emanate from these are along the lines of, 'In the real world you never get the right people, therefore one has to make do with lower grades, and one needs procedures to control them'

My first response to this is to say that if your staff are so lacking in intellect that they need procedure to tell them when and how to go to the toilet you are doomed anyway (I exaggerate, but have you seen a PRINCE manual?). However, the criticism does have some basis in truth. I therefore, as part of my social and human factors based methodology, divide my life cycle into two phases:

1. Phase A is where a small, creative communicative team of IT and domain specialists scopes the system, elicits the requirements at a fundamental abstract level, anticipates future changes, and designs and codes the Kernel of the system. It also designs the test philosophy and top level tests.

2. Phase B is where a larger team composed of technicians grinds out all the code, performs the testing and copes with cosmetic changes. This phase will merge with what is conventionally known as the maintenance phase.

Obviously there is a need for the continuity of some staff between the two phases. However, the approach would have the advantage that scarce creative people are concentrated where they are really required.

Some may argue that safety-critical systems are not amenable to this approach. Presuming that their arguments are based on the hypothesis that the only difference between conventional IT systems and safety-critical ones, particularly with respect to software, is that the latter have to be demonstrated to be safe in advance of operation, one could easily refute this. Firstly, if the social approach produces the better product and this is accepted, there is nothing within it that precludes the demonstration of its safety-producing properties. However, I would concede that good social features of the production process are not as easily demonstrable as mountains of paper indicating that various techno-processes have been gone through. Secondly, there is, in my opinion, too much emphasis on the assessment of the production process rather than the resultant product. Whilst it can be argued that one is more likely to get a good (in this instance, safe) product from a good (safety cognisant) process rather than a process that is less good, one cannot guarantee it. We really need better techniques for assessing the safety of the product rather than the process which produces it.

In fact human factors and social factors may be the best guarantee of safety. Some argue the best guarantee of the safety of software may well be our scepticism of the inherent safety of software, and this is why such systems have not killed many people yet.[1] An important feature of my social method when applied to safety-critical systems therefore comprises the following notions:-

At the outset of the project one should ask oneself about the safety relevance of the proposed project, including whether legislation or standards deem it safety-related or safety-critical;

Continually during the conversion of requirements into a system, those involved should ask the question, 'Have I introduced any feature that compromises safety?'

If the answer to either question indicates that safety is compromised, the necessary corrective action should be taken as part of a safety culture.

One need only observe the behaviour of the average motorist to see the difficulties of introducing a safety culture, in a particular domain. However it may well be that the social methodology would help in this respect, based on the two assumptions viz:-

Most safety-threatening and safety-providing issues will arise and be dealt with in Phase A;

The individuals employed during Phase A would be more likely to have the creativity to perceive and handle safety threats. This is of course debatable unless the ivory towerness of Phase A was leavened with practicality.

Another safety issue is: would the social methodology apply to both programmable electronic systems (pes) and programmable logical controllers (plc)? I believe it would. However, in passing, I would counsel against the popular perception that the latter are a sub-set of the former. They emanate from different cultures and have a significant life-cycle difference in that the 'should it be hardware or software' design decision is made at a much lower level in the case of plc compared to pes.

Promoters of a social methodology are of course most likely to be criticised by those that are techno-centred. I would not therefore like to ignore technical issues and fall into the reverse of the trap that they fall into.

The problem is what particular techniques and tools should influence the social method. At a recent Safety-Critical Systems Club meeting, I was spellbound by successive talks on Static Testing, Fagan's Inspection, Validation and Formal Methods and deeply impressed by their merits. The problem is that there is no quantitative feel to enable one to compare their merits or to determine where each should be applied or how they could be used together.

Take, for example, the most frequently advanced technology solution to the problem of safe software - formal methods. If we ignore the problems of 'are the requirements right anyway?' and their incomprehensibility to the proverbial man on the Clapham Omnibus, it would be useful to know the maximum size of system they have fully been used on. Top of the Pops at the moment is the 'Paris Metro' with 20,000 lines of code and 7,000 proof arguments.[2] Other railway examples cited lie within this and exhibit a similar 3:1 ratio for lines of code to proof arguments. However, it could be argued that railways are not a severe test, given the maturity of the domain in safety terms and the fact that two metal rails constrain the complexity of the problem. I am currently investigating a larger, more complex example from another domain [3] but at the moment the Paris Metro represents the upper limit of system where their usefulness has been validated. But where do you see such information conveniently published?

As you will gather, my own inclinations on the best process for the production of safe systems is based on a social methodology. This utilises techniques and tools based on their appropriateness to the domain, the understanding of them by the persons involved, their compatibility one with another, and of course their demonstrated ability to yield quality or productivity benefits.

After all, I know a lot of working software that has been produced without specific techniques and tools but none that has been produced without people.

References

1 MacKenzie D: Computer related accidental death: an empirical exploration. Science and Public Policy, August 1994.
2 Dehbonci B and Mejial F: Forma Methods in the Railways Signalling Industry. Proceedings of Formal Methods Europe, Oct. 1994.
3 Using Formal Methods to Develop an ATC Information - System. IEEE Software, Vol. 13, No. 2, March 1996.

The Need for New Paradigms in Safety Engineering

Nancy G. Leveson

MIT

First published in Safety Systems 17-2, January 2008.

Most of the safety engineering techniques and tools we use today were originally created for first mechanical and later electro-mechanical systems. They rest on models of accident causation that were appropriate for those types of systems, but not the majority of the systems we are building today. After computers and other new technology became important in most new systems, the primary approach to handling safety in these systems was to try to extend the traditional techniques and tools to include software. We have now attempted that for at least three decades with little real success. I believe that it is time to consider that this approach may not lead to great success and that something else is needed [6,8].

The problem appears to be that software and digital devices operate (and, in particular, fail) very differently than analog devices. In addition, software allows us to increase the complexity of the systems we build (in particular, interactive complexity and coupling) such that new types of accidents are occurring that do not fit the traditional accident causation model. So called 'system accidents' arise not from the failure of individual system components, but from dysfunctional interactions among system components, none of which may have failed. A model of accident causation and the engineering techniques built on it that consider only component failures will miss system accidents, which are the most common software-related accidents. In addition, the role of human operators is changing from direct control to supervisory positions involving sophisticated decision-making. Once again, the types of mistakes humans are making are different and are not readily explained or handled by the traditional chain-of-failure-events models. Finally, there is more widespread recognition of the importance of management, organizational, and cultural factors in accidents and safety: the traditional models, never derived to handle these factors, do so poorly if at all.

I believe that to make significant progress in safety engineering, we need to rethink the old models and create new accident causality models and engineering techniques and tools based on them that include not only the old accident causes but also the new types of accidents and accident causality factors. In this article, I suggest one such model and some tools based on it [5], but it is not the only such model possible and other tools and techniques might be built on it or on other models. Our new model is based on system theory (rather than the reliability theory of the traditional models) and our experience with it has shown that it allows much more powerful accident analysis and root cause analysis [7], hazard analysis [1], design-for-safety techniques [1], and general approaches to risk management in complex, socio-technical systems [3,5].

STAMP: An accident causality model based on system theory

Traditional accident causation models explain accidents in terms of a chain of events that leads up to the accident. The relationships assumed between events in the chain are direct and relatively simple. Using this model of causation, the most appropriate approaches to preventing accidents is to somehow 'break the chain' by either preventing an event or by adding additional 'and' gates in the chain to make the occurrence of the accident chain less likely. As the events included almost always involve component failures or human errors, the primary mechanism for increasing safety is to make the individual components more reliable or failure free. Such models are limited in their ability to handle accidents in complex systems, organizational and managerial (social and cultural) factors in accidents, human error, and the systemic causes of the events.

For the past five years, the author has been developing a new, more comprehensive model of accident causation, called STAMP (System- Theoretic Accident Model and Processes), that includes the old models but can better handle the levels of complexity and technical innovation in today's systems [4,6]. STAMP is based on systems theory and includes non-linear, indirect, and feedback relationships among events. Accidents are explained as resulting not simply from system component failures but also from interactions among system components (both physical and social) that violate system safety constraints. System safety is treated as a control problem (rather than simply a failure problem): accidents occur when component failures, external disturbances, and/ or dysfunctional interactions among system components are not handled adequately by the control system (where the controls may be managerial, organizational, physical, operational, or manufacturing) such that required safety constraints on system behavior are violated. Major accidents rarely have a single root cause but result from an adaptive feedback function that fails to maintain safety as performance changes over time to meet a complex and changing set of goals and values. The accident or loss itself results not simply from component failure or human error (which are symptoms rather than root causes) but from the inadequate control (i.e., enforcement) of safety-related constraints on the development, design, construction, and operation of the entire socio-technical system.

In STAMP, the events that precede a loss reflect the effects of dysfunctional interactions and inadequate enforcement of safety constraints, but the inadequate control itself is only indirectly reflected by the events—the events are the result of the inadequate control. The control structure itself, therefore, must be carefully designed and evaluated to ensure that the controls are adequate to maintain the constraints on behavior necessary to control risk. Figure 1 shows an example control structure for a typical U.S. regulated agency, such as aircraft. Each industry and company will, of course, have its own control structure.

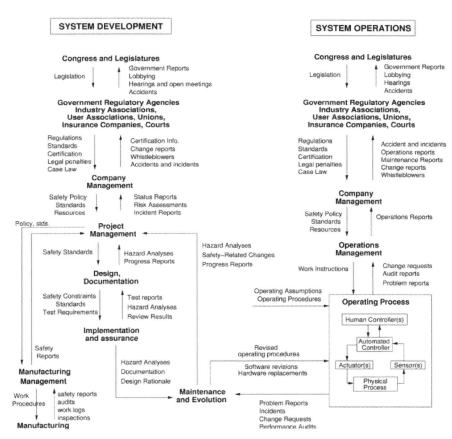

Fig. 1: Example Model of Socio-Technical Control

Note that the use of the term 'control' does not imply a strict military command and control structure. Behavior is controlled not only by engineered systems and direct management intervention, but also indirectly by policies, procedures, shared values, and other aspects of the organizational culture. All behavior is influenced and at least partially 'controlled' by the social and organizational context in which the behavior occurs. Engineering this context can be an effective way of creating and changing a safety culture.

A major advantage of this approach is that it can handle very complex systems— aspects of STAMP have been successfully applied to complex government projects such as the U.S. Missile Defense System and the NASA manned space program. And it is able to handle both the technical and the social (organizational and cultural) aspects of accident understanding and prevention.

Technical tools based on STAMP, such as new hazard analysis techniques and approaches to design for safety, can be used on the physical system design alone, as is usually done today by safety engineers. The approach involves identifying the physical constraints that must be enforced and ensuring that the technical and social system adequately enforces them through various types of controls. It also identifies the required process model (mental model if the controller is a human) that the controller needs in order to provide adequate control and thus the information required in that process or mental model.

But STAMP can go beyond physical system design. New approaches to organizational risk analysis based on STAMP involve creating a model of the social and organizational control structure and identifying the safety constraints each component is responsible for maintaining, a model of the social dynamics and pressures that can lead to degradation of this structure over time, process models representing the view of the process by those controlling it, and a model of the cultural and political context in which decision-making occurs.

For both technical system hazard analysis and organizational risk analysis, we apply a set of factors we have identified that can lead to violation of safety constraints, such as inadequate feedback to maintain accurate mental (process) models. These factors are derived from basic control theory. For hazard analysis, the information that results can be used to guide the design of the system in a 'safety-driven' process that builds safety in from the beginning of the design activity. For organizational risk analysis, the information resulting from the modeling and analysis effort can be used to assess the risk in both the current organizational culture and structure and in potential changes, to devise policies and changes that can decrease risk and evaluate their implications with respect to other important goals, and to create metrics and other performance measures (leading indicators) to identify when risk is increasing to unacceptable levels

Conclusions

STAMP is not the only possible expanded model of accident causation that could be devised. The purpose of this article is not to sell STAMP, but to encourage those working in this field to expand beyond the techniques and models created for simple electro-mechanical systems whose underlying assumptions no longer match the majority of the systems we are building today.

By creating new models, we will be able to provide much more powerful safety engineering techniques and tools. This hypothesis is supported by our experience with STAMP and the new procedures, techniques, and tools built on top of this new accident causation model. For example, it has been used successfully at the physical (technical) system level to provide a non-advocate safety assessment of the new U.S. Missile Defense System, an extremely complex 'system of systems' made up of both components that have existed for decades (e.g., early warning systems) and more recent or new components (e.g., radar systems and interceptors). Traditional hazard analysis and safety assessment techniques were ruled out for this goal as impractical. Instead, a new STAMP-based hazard analysis technique [1], called (STamP Analysis) was used. The assessment was successfully completed on the integrated system; in fact, so many potential paths to inadvertent launch (the first hazard analyzed) were identified by the STPA analysis that deployment and testing of the missile defense system was delayed for six months beyond the planned date.

Demonstrations of STAMP-based, safety-driven design of complex systems, such as spacecraft, have been successfully completed that show how to use preliminary hazard analysis (starting in the very early concept formation stages of system design) to inform the design decisions as they are being made so as to eliminate much of the rework that results from hazard analysis that occurs after the design has been created [2]. To accomplish this goal, more powerful hazard analysis techniques are needed that go beyond simple definitions of failure as the cause of accidents and that do not rely on already having a completed design [1].

We have also completed demonstrations of applying STAMP to organizational and cultural risk analysis and risk management in the manned space program, both the current Space Shuttle operations program [5] and the development of the new U.S. manned space exploration program to return to the moon and then go to Mars [3]. In both analyses, our models start with Congress and the White House and continue down through the NASA management structure to the engineering project offices and the actual operations (in the case of the Space Shuttle) and system development (for the Exploration Systems Mission). In the operational risk analysis of the current Space Shuttle program, we identified system-level requirements to reduce poor engineering and management decision-making leading to an accident, identified gaps and omissions in the operational program design, and performed a rigorous risk analysis to evaluate proposed policy and structure changes and to identify leading indicators and metrics of migration toward states of unacceptable risk over time. In the Exploration Systems risk management demonstration, we showed how STAMP-based analysis could be used to assist in understanding the tradeoffs between schedule, budget, performance, and safety risks.

Most recently we have been exploring the limits of this new approach and applying it to a large variety of risk management problems, including safety in pharmaceutical testing, healthcare, the process industry, and the air transportation system, as well as to corporate fraud and security of national infrastructure systems.

References

1. Nicolas Dulac and Nancy Leveson, An Approach to Design for Safety in Complex Systems, International Conference on System Engineering (INCOSE), Toulouse, June 2004.
2. Nicolas Dulac and Nancy Leveson. Incorporating Safety into Early System Architecture Trade Studies, International Conference of the System Safety Society, August 2005.
3. Nicolas Dulac, Brandon Owens, Nancy Leveson, Betty Barrett, John Carroll, Joel Cutcher-Gershenfeld, Stephen Friedenthal, Joseph Laracy and Joel Sussman, Demonstration of a Powerful New Approach to Risk Analysis for NASA Project Constellation, CSRL Final Project Report, March 2007 (can be downloaded from http://sunnyday.mit.edu/papers.html)
4. Nancy Leveson, A New Accident Model for Engineering Safer Systems, Safety Science, 42(4), April 2004, pp. 237–270.
5. Nancy Leveson, Nicolas Dulac, Betty Barrett, John Carroll, Joel Cutcher- Gershenfeld, Stephen Friedenthal, Risk Analysis of NASA Independent Technical Authority, CSRL Final Report, June 2005 (can be found on http://sunnyday.mit.edu/papers.html)
6. Nancy Leveson, System Safety Engineering: Back to the Future, unfinished manuscript, http://sunnyday.mit.edu/book2.html
7. Nancy Leveson, Mirna Daouk, Nicolas Dulac and Karen Marais, Applying STAMP in Accident Analysis, Second Workshop on the Investigation and Reporting of Accidents, Williamsburg, September 2003.
8. Karen Marais, Nicolas Dulac and Nancy Leveson, Beyond Normal Accidents and High Reliability Organizations: The Need for an Alternative Approach to Safety in Complex Systems, ESD Symposium, March 2004.

Nancy Leveson is Professor of Aeronautics and Astronautics and also Professor of Engineering Systems at MIT. She is an elected member of the National Academy of Engineering (NAE). Prof. Leveson conducts research on the topics of system safety, software safety, software and system engineering, and human-computer interaction. In 1999, she received the ACM Allen Newell Award for outstanding computer science research and in 1995 the AIAA Information Systems Award for 'developing the field of software safety and for promoting responsible software and system engineering practices where life and property are at stake.' In 2005 she received the ACM Sigsoft Outstanding Research Award. She has published over 200 research papers and is author of a book, 'Safeware: System Safety and Computers' published by Addison-Wesley.

Medical radiological incidents: the human element in complex systems

Chris Moore

> Christie Hospital
>
> Manchester

First published in Safety Systems 5-2, January 1996

1 Introduction

Mix medical risk with radiation damage and the resultant cocktail induces an overtly emotional response. Radiation induced damage is used to cure cancer but the art of medical practice is to spare healthy tissues that ordinarily look too close to the diseased tissue for comfort. It is a balancing act based on empirically derived biological probabilities of tumour control and normal tissue complication, and is performed successfully on thousands of unique individuals each year. This is radiotherapy where under-dosing, as much as overdosing, can lead to treatment failures. Why then, after decades of refinement, should the introduction of new technology, to improve treatment quality and widen its availability, focus attention on disastrous outcomes? Probably because through such disasters it is becoming increasingly obvious that mere mortals are the really weak link in the complex chain.

1.1 The Spectacular and the Insidious

Radiotherapy is a frequent procedure and the consequences of its failure can be terminal. Thus the risks might be considered relatively high albeit well understood. What is not widely understood is the protracted nature of radiotherapy, and that it takes many years of follow-up studies to assess treatment effectiveness. Technical failure adds to inherent risk, becoming apparent on a time scale ranging from days to years. Insidious failure is in fact much more of a disaster that overt catastrophe since thousands are affected. Unfortunately it is less immediate, less spectacular, more difficult to identify and to eliminate, and so is consigned to statistical records. In contrast the spectacular deaths of individuals remain tragedies sharply focused in our minds. Thus Bhopal and the deaths of thousands has been compared to just six from the Therac-25 accidents. Yet embedded in the Therac story is the possibility of a far greater tragedy that goes generally unnoticed.

1.2 Complex now, more so Tomorrow

Today's radiotherapy is already scientifically and technically complex. Underlying computerised treatment planning are advanced physical dosimetric models. Graphical simulation using these models precedes interpretation of the plans for a given patient. This is followed by the delivery of treatment, using high-energy, computer-controlled irradiation utilities. And, last but not least, there is monitoring and verification throughout the delivery of treatment, discretely phased over days or weeks. Hence inadequate computer operation at any stage of the radiotherapy process can lead to catastrophe.

The next phase of computerisation is about to take the use of radiotherapy treatment machines from a stand-alone to a networked environment managed by a central server. Staff will require a lot more than end-user abilities to control the delivery of treatment. The demands for skills and knowledge of underlying processes will increase as routine treatments give way to patient-tailored treatments. Yet the idea that automation of the process allows abdication of responsibility persists, particularly at managerial level where an opportunity for technical de-skilling is seen. Fewer in the manufacturer-user chain will have a complete understanding of what they are doing and the risks being taken. In this context accidents described below already provide us with lessons we have failed to learn from. As the public at large increasingly believe that mortality is optional, there is bound to be an imposed move towards the development of safer systems in a radically re-structured medical practice.

2 Human Error in Computerised Treatment Planning: The Stoke Incident

A protracted incident was discovered at the North Staffordshire Royal Infirmary (NSRI) in 1991. The incident started shortly after the introduction of new computerised treatment planning hardware in 1982, continuing until its discovery nearly ten years later. By that time incorrect practice in computerised treatment planning had already led to the 5-30% under-dosing of 989 patients with consequent increased risk of disease recurrence and mortality.

Prior to computerised planning at NSRI, so-called isocentric treatment plans could only be produced manually, so complete irradiation plans were impractical. Radiographic staff were aware of the need for a radiation source-to-skin distance correction, and they included this in such plans. The first case of computerised isocentric treatment planning was performed by one radiographer responsible for planning, one radiographer responsible for treatment delivery, and one physicist whose role was purely advisory. The treatment delivery radiographer was not aware that the computer algorithm for this type of treatment already included the necessary distance corrections, and so followed the old practice of including the correction - a second time. There was in fact an uneasy relationship between the physicist and other groups, a matter which had been raised with the radiographers, clinicians and hospital management and elsewhere between 1977 and 1991, and was due to unclear allocation of responsibilities. An inquiry report stated:

"Supervision of planning and treatment rested with people who had not received professional education and training to graduate or postgraduate level in physics and were less well equipped therefore to understand as fully as Radiotherapy Physicists the process by which the concepts of radiation physics are harnessed to clinical procedure. We encountered instances of inability among radiographers to recognise isocentric plans for what they are Had it been agreed at any time between 1982 and 1991 that such a particular supervisory role should be given instead to the Principal Radiotherapy Physicist, as the senior person with professional understanding of the entire process, we would have expected the fault to have been detected long before 1991."

Non-physicists are still used for planning in many centres. Hence it is still widely assumed that what appears on a treatment planning computer monitor is 'correct'. Following a report on an earlier 22-patient incident in Exeter, not involving computer use, government ministers requested advice on quality assurance in radiotherapy. The government made funds available for two pilot centres to put into practice the philosophy of BS5750, one centre being the Christie Hospital in Manchester. Following Stoke, there appear to have been no recommendations regarding the development and application of treatment-planning software. Scientific staff remain scarce and most end users have little knowledge of computers or software.

3 Human Error in Treatment Delivery

Treatment delivery errors can be catastrophic. Incidents in Spain and the USA show how the same inter-professional problems and lack of knowledge by those clinically directing treatments, evident from the Stoke incident, can contribute to fatalities.

3.1 The Zaragoza Incident

On December 5th 1990, an energy-selectable electron treatment unit at the University Hospital of Zaragoza, Spain, became faulty. An engineer, on site from the manufacturer, briefly examined the unit but was unable to identify what turned out to be a recurrence of a known problem. On December 7th the engineer performed a 'quick fix', overriding the security systems to get the machine back into some form of operation for the following Monday, December 10th. Prior to use a radiographer questioned the indication of 36 MeV electron output on a meter, with the machine still off. The meter was assumed to be jammed and so the treatment machine went into clinical use with radiographers setting normal treatment values on the machine's control console.

No report of the problem, or 'repair', was made, let alone given to the local physicists. Due to a lack of serviceable equipment, below even that recommended by Spanish protocols, it was another 10 days before the scheduled monthly dosimetry checks were performed by physicists. In fact physicists were not regarded as essential, since they have no legal role and it is the clinical staff who receive licensing for the running of irradiation installations. On December 20th the physicists were told about the jammed meter and the fixed 36 MeV energy indication. The physicists found that regardless of the selection of electron energy at 7, 10 or 13 MeV, the true energy was always 36 MeV. Consequently there were massive overdoses at these lower energies of up to 15 times those intended. Of 27 patients treated, 10 have died as a result of the error.

3.2 The Therac Incidents

The first incident in Marietta, Georgia, in 1985, falls neatly into the middle of the 'Stoke decade' and once again highlights the same human failings.

A patient overdosed, protests ignored by the treatment radiographer, schedules for continued treatment by clinicians despite obvious clinical damage, until after two weeks a physicist noticed paired radiation reaction marks on the patient and correctly deduced the nature of the overdose.

In Hamilton, Ontario, in 1985, an operator tried 4 times to skip over an indicated fault before calling for technical help, to no avail as it turns out.

In the same year at Yakima, Washington, marks on a patient went 'unexplained'; 1986 saw another incident, at Tyler in Texas, where a machine dose error was caused and subsequently skipped over because of operational familiarity. This treatment was delivered (twice in fact) without any working video or audio contact to the patient, who was in a sealed room and aware that something was malfunctioning. A second fatal incident occurred a few weeks later, in almost the same way, involving the same operator.

A University of Chicago therapy physicist was able to deduce the user interaction sequence causing the Therac accidents. Catastrophic shutdowns of an older teaching facility, equipped with independent hardware interlocks as well as software controls, occurred more frequently with new intake students who were undisciplined with their responses to control console requests. Errors subsequently disappeared for months. The inference is that bad practices seen in new students re-appear at a later stage with machine familiarity. Between times, the users have a healthy respect for the inherent dangers of the therapy process and interact with appropriate caution.

Since the earlier Therac machines had hardware interlocks preventing overdose to patients, the use of software to replace them was both unnecessary and retrograde. The fact that subsequent reports highlighted less than ideal software engineering practices should be no surprise. Treatment machine manufacturers' core business is not software development.

The reliance on software for safe operation was clearly a basic mistake. But reports barely mention the need for education, additional to training, for treatment machine operators. Users should understand the sequence of events they are initiating; so too should the software designers.

Is there a hidden tragedy to the Therac incidents? Discussions at user meetings following the Therac accidents revealed other problems leading to 10 to 30% under-dosing when either electron or X-rays were used. This suggests that a far greater disaster may have occurred, along the lines of the Stoke incident! Yet there appears to have been no publicised analysis of the problem in relationship to software flaws. The reason for this may be cultural, where perhaps in the USA under-dosing does not place a hospital or manufacturer at serious fault, since it is not radiation damage but the disease that takes its natural course in harming the patient. In reality, of course, it is the failure to treat correctly that is the cause of harm.

4 Conclusion

If incidents like those in Stoke, Zaragoza and North America tell us anything, beyond the mechanical, electronic or software reasons for fatalities, it is that the management of hospitals and the organisation of professionals within them continue to reflect a past era of low-technology. Technology solutions to technology problems will have a minor impact on radiotherapy incidents. Technology solutions to human problems may fare much better.

The ALARP Principle:
Legal and Historical Background

Felix Redmill

This is the first of three articles by Felix Redmill on the subject of ALARP. All three articles are presented together and counted as one in this volume.

First published in Safety Systems 18-2, January 2009

Introduction

The UK's Health and Safety at Work Etc. Act 1974 [HSW Act 1974] requires that risks imposed on others (employees and the public at large) should be reduced 'so far as is reasonably practicable' (SFAIRP), but it offers no guidance on how to determine what is SFAIRP in any given circumstance. It is the Health and Safety Executive's ALARP (as low as reasonably practicable) Principle that provides a model for doing this.

A previous article [Redmill 2008] offered an explanation of the ALARP model, and a subsequent one will explore its use in practice. This article reviews its origins and development and, briefly, its application in law.

'Reasonableness' in the English Common Law

Some legal offences are defined by prescription, for example a defined threshold (such as a speed limit), which it is illegal to exceed. But in the majority of cases there is no definitive prescription, the boundary between criminal and acceptable behaviour depends on the circumstances, and there may be plausible arguments on both sides – for both guilt and innocence. Traditionally in such cases, under the English Common Law, judges attempted to determine what is 'reasonable' in the circumstances. And their way of doing this was to appeal to the concept of 'the reasonable man', asking, in effect: How would a reasonable person determine this case? In order to suggest that this notional appeal was not to someone above the status of the appellants, but, in fact to an 'ordinary person' and, by implication, their peer, judges often referred to 'the man on the Clapham omnibus'.

The tradition of judges determining what is 'reasonable' – in a manner understandable and acceptable to those engaged in the case – led to both legal precedents and statutes that hinge on the interpretation of that word. For example, it is permitted in law to use reasonable force to repel an assailant, or a burglar, and in such cases the criterion of reasonableness is proportionality to the threat imposed by the aggressor.

In the Corporate Manslaughter and Corporate Homicide Act 2007, a breach of a duty of care is defined as a 'gross' breach 'if the conduct ... falls far below what can reasonably be expected of the organisation in the circumstances.' Responsibilities of the court must therefore be, first to understand the circumstances of the case, second, to determine what is or would be reasonable, and, on the results of these findings, to pass judgement.

Thus, reasonableness has traditionally been, and remains, a criterion of judgement in courts of law. At the same time, it is a variable whose meaning must be determined against criteria extracted from the circumstances of the case.

In the context of risk, what is reasonable depends, among other variables, on what is to be gained by taking the risk. But because the potential benefits are usually not equally distributed between those creating a risk and those having it imposed on them, disagreements are likely to arise and to come to court for resolution.

Edwards vs. The National Coal Board

The obligation to do what is 'reasonably practicable' to reduce risk was enunciated in a case in 1949 which gave definition to the way of thinking that is now referred to as 'riskbased' and which eventually resulted in the development of the ALARP Principle. The plaintiff, Edwards, alleged negligence by the National Coal Board, claiming that organisations have a duty of care to their employees and that they should do all that is 'reasonably practicable' to protect them from harm. In an attempt to explain how 'reasonable practicability' should be determined, the judge said: 'a computation must be made in which the quantum of risk is placed on one scale and the sacrifice, whether in money, time or trouble, involved in the measures necessary to avert the risk is placed in the other, and that, if it be shown that there is a gross disproportion between them, the risk being insignificant in relation to the sacrifice, the person upon whom the duty is laid discharges the burden of proving that compliance was not reasonably practicable.' [Edwards vs. The National Coal Board 1949]

The Robens Report

The next significant development in the evolution of ALARP occurred in the early 1970s, when the UK's government initiated a fundamental review of the regulation and management of occupational risks. The resulting 'Robens Report' [Robens 1972], named after the committee's chairman, recognised that the extent of occupational risk was such that health, safety and welfare at work could not be ensured by an ever-expanding body of legal regulations enforced by an ever-increasing army of inspectors. It therefore recommended that the primary responsibility for ensuring health and safety should lie with those who create risks and those who work with them. The report further recommended that to create such an environment the law should provide a statement of principles and definitions of duties of general application, with regulations setting more specific goals and standards.

The Robens Report's recommendations for both legislation and regulation were implemented shortly afterwards. The Health and Safety at Work Etc. Act 1974 [HSW Act 1974] was introduced. It contained the basic legislation and it allocated the responsibility for the regulation of health and safety risks to the Health and Safety Commission (HSC) and the Health and Safety Executive (HSE). What is to be done about any particular risk is defined in the Act, not absolutely but in terms of proportionality between the costs of reducing the risk and the benefits to be gained by doing so. According to the Act, risks should be reduced 'so far as is reasonably practicable' (SFAIRP), which demands no more than is reasonable and possible to deliver, with the onus being on the imposer of a risk to justify its residual value by demonstrating that further reduction would not be reasonably practicable. The principle of reasonable practicability, as well as that of gross disproportion, formerly enunciated in case law [Edwards vs. The National Coal Board 1949], thus became enshrined in statute.

It should be noted that the Act does not offer a justification for the acceptance of huge risks on the grounds that their reduction would be prohibitively expensive. Yet, unlike the ALARP model [HSE 1992, Redmill 2008], which includes a 'limit of tolerability' threshold (see Figure 1), it does not explicitly state that there is a threshold beyond which risk should not be accepted. However, accepting a very high risk demands exceedingly strong justification and, in attempting to comply with ALARP, the assumption of a 'limit of tolerability' threshold is a wise precaution. If its value on the risk scale has not been determined on an industry-sector basis, it must be determined for the particular case in question, and demonstration of good judgement in this could subsequently be evidentially useful in a court of law.

Publication of ALARP

In 1987, a Public Inquiry into the proposed Sizewell B nuclear power station [Layfield 1987] recommended that the HSE should formulate and publish 'guidelines on the tolerable levels of individual and social risk to workers and the public from nuclear power stations.' Not only was the concept of such guidelines on tolerable risk new, so was Layfield's recognition that risk analysis involves both scientific assessment and social values. One of his recommendations was that 'the opinion of the public should underlie the evaluation of risk; there is at present insufficient public information to allow understanding of the basis for the regulation of nuclear safety.' The result was the publication, by the HSE, of the ALARP Principle as guidance to the nuclear industry on the approach to reducing risks SFAIRP [HSE 1988, 1992].

Since the 1980s the HSE has not only given a great deal of advice to risk creators about their duties of care, but also taken many initiatives to inform the public both of what risk management should entail and how the HSE itself goes about its duty of assessment and regulation (in particular [HSE 2001]).

Since the ALARP Principle came into being in 1988, its application has been extended by the HSE from its original context in the nuclear industry. It is now the model on which regulation across all industry sectors is based, and it sets the terms of reference for making risk tolerability decisions by all who would impose risks on others. Its precepts, and the model (Figure 1) that represents it, are not hard to understand. However, though it may appear simple, it is not trivial to apply in practice, for it depends heavily on interpretations, in the circumstances, of what is reasonable and what is practicable.

Self-regulation

The system defined by ALARP, of risk determination, assessment, reduction and management by those who create risks, amounts to self-regulation – as had been recommended by the Robens Report. Thus, the Act is not prescriptive but goal-based. The creator of a risk must set and justify the risk-reduction goals and demonstrate that they have been achieved; an independent assessor must assess and approve (or reject) the goals and verify their attainment. The regulator, if involved, has the right, on behalf of the public and the law, to assess the discharge of both the risk-creator's and the assessor's responsibilities.

Practicability not Practicality

The HSE has explained that the word 'practicable' was deliberately chosen over 'practical', as the latter would imply that a risk should be reduced as far as physically possible and that what is intended is that the cost of reducing the risk should be taken into account. Borrowing heavily from the words of the judge in the 1949 case of Edwards vs. the National Coal Board, the HSE defined 'reasonably practicable' in the following terms: 'Legally speaking, this means that unless the expense undertaken is in gross disproportion to the risk, the employer must undertake that expense' (HSE 1992, Para 32). However, recognising that there will always be some risk and that the question is, how much, the HSE goes on to say that 'reasonably practicable' also means that there is a point beyond which the regulator should not press the operator to go – 'but that they must err on the side of safety.'

Application in Law

The Health and Safety at Work Act is unusual in English law in that it places the onus on the defendant to demonstrate innocence – by proving that everything was done, so far as was reasonably practicable, to ensure safety by the reduction of risk. When a case is brought to court, the things that have been done – or not done – by the accused are matters of fact. Whether it was reasonable to have done them – or not done them – are matters of judgement, to be made by the court.

When risk is involved, reasonableness depends on what it was practicable to do to reduce the risk, or risks. And practicability comprises three factors: what it was possible to do and, for each possibility, the costs of doing it, and the benefits gained. As the costs become greater, there may come a point at which they become 'grossly disproportionate' to the benefits gained, at which point a defendant demonstrates that further reduction would not be reasonably practicable and that the risk in question is ALARP – as long as the residual value of the risk is not greater than the limit of tolerability threshold (see Figure 1).

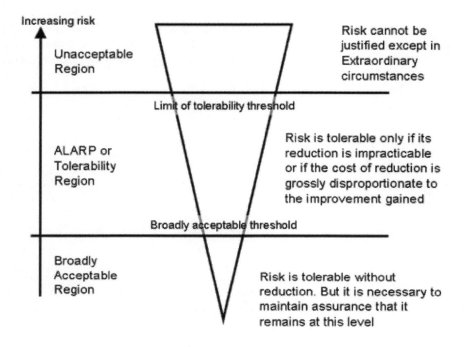

Fig. 1. The ALARP Model

The point of gross disproportionality is different for the different courses of action to reduce a risk, so a court would need to be convinced that all avenues had been explored before a decision was made on what should be done. Further, as the point of gross disproportion is a matter of judgement, we cannot be certain that our determination of it now will find favour with a court later; therefore, it is wise to err on the side of caution, particularly when the risk is high. (The HSE suggests that 'a disproportion factor (DF) of more than 10 is unlikely' and that 'the duty holder would have to justify use of a smaller DF' [HSE 2008].)

In addition, although the ALARP model represents a sensible way of attempting to comply with the law, it is not a part of the law. The values of risk at its two thresholds are not defined in law. Indeed, the thresholds themselves do not appear in the Act; they exist only in the HSE's model in order to emphasise the fact that different types of decision-making occur at different levels of risk. Even then, they are matters of calibration. The HSE's recommendations for calibration in the nuclear industry may offer sensible values for the thresholds, but there is no guarantee that a court will take these to be definitive.

Moreover, where the principle of reasonableness is invoked, no case provides an absolute precedent for others; each must be judged on its own circumstances. Suppose, for example, that a drug has a side effect that causes five deaths per million of those who take it. If the drug is efficacious and saves many other lives, and the cost of reducing the risk is significant, a court may hold that it is reasonable for the manufacturer not to reduce it further. But suppose a vegetable supplier sprays cabbages with an additive that makes them look fresher but which causes the deaths of five per million cabbage eaters. Would a court accept that there is no need for the supplier to reduce such a risk on the grounds that it is too expensive to do so? It is at least conceivable that the court would find that there is no justification for the risk, small as it is, and that it should have been avoided altogether. It is also possible to imagine a situation in which it is deemed unreasonable to reduce a risk that lies in the model's tolerability region – if the method of reduction has an unintended consequence that throws up a new and greater risk. The ALARP model is a tool that should be used with understanding and discretion.

Summary

The ALARP Principle arose out of legal concepts, precedents and requirements. It is not a technical concept, and its purpose is not to meet engineering requirements. Rather, it is an attempt to bridge the gap – sometimes perceived as an abyss – between those responsible for the safety of technological systems and the requirements placed on them by UK law. It is a framework to facilitate good judgement on safety-related risk-reducing actions (technical or otherwise) in order to comply with the law. In the matter of interpreting the law, it is a bridge from the legal to the engineering; in fulfilling the requirement for risk reduction to be SFAIRP, it is a bridge from the engineering to the legal. This article traces its evolution and shows its relation to the law.

Felix Redmill provides training in safety engineering and management principles, risk, and project management.

References

Edwards vs. The National Coal Board [1949]. 1 All ER 743 HSE [1988]. Health and Safety Executive: The Tolerability of Risk from Nuclear Power Stations. Discussion Document, HMSO, London

HSE [1992]. Health and Safety Executive: The Tolerability of Risk from Nuclear Power Stations. HMSO, London

HSE [2001]. Health and Safety Executive: Reducing Risks, Protecting People. HSE Books.

HSE [2008]. Cost Benefit Analysis (CBA) Checklist. http://www. hse.gov.uk/risk/theory/alarp1.htm (After update on 08.04.08)

HSW Act [1974]. The Health and Safety at Work Etc. Act. HMSO, London
Layfield, Sir Frank [1987]. Sizewell B Public Inquiry Report. HMSO, London
Redmill, Felix [2008]. Overview of the ALARP Principle. Safety Systems, Vol. 17, No. 3, May 2008
Robens, Lord [1972]. Safety and Health at Work: Report of the Committee, Cmnd 5034. HMSO, London

Making ALARP Decisions

Felix Redmill

This is the second of three articles by Felix Redmill on the subject of ALARP. All three articles are presented together and counted as one in this volume.

First published in Safety Systems 19-1, September 2009.

Introduction

The UK Health and Safety Executive's (HSE) ALARP (As Low As Reasonably Practicable) Principle is intended to provide guidance on making risk-tolerability decisions in compliance with the legal requirement to reduce risks 'so far as is reasonably practicable' (SFAIRP). But, though the ALARP Principle - represented by its model (see Figure 1) - is itself not difficult to understand, applying it is not straightforward. The triangular balancing of risks, the costs of reducing them, and the potential benefits of doing so is a source of debate, dispute, and even despair for many system safety engineers and managers. We can never be certain that our judgement at the time of acting will find favour with a court, sitting later, with retrospective knowledge

© Felix Redmill 2009.
Published by the Safety-Critical Systems Club. All Rights Reserved

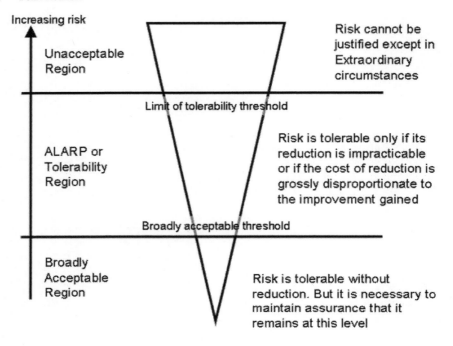

Fig. 1. The ALARP Model

Three previous articles have explained the Principle itself, sketched its origins and history, and outlined the preliminary processes essential to its effective application [Redmill 2008, 2009a, 2009b]. This article addresses the use of the ALARP model for making risk-tolerability decisions. Drawing on HSE guidance literature, it presents a route through the process by means of the flowcharts of Figures 2 and 3. The preliminary processes previously explained include the estimation of risk values, so at the point of entry (B) to Figure 2, it is already known where on the ALARP model (Figure 1) each risk is placed.

Making ALARP Decisions 41

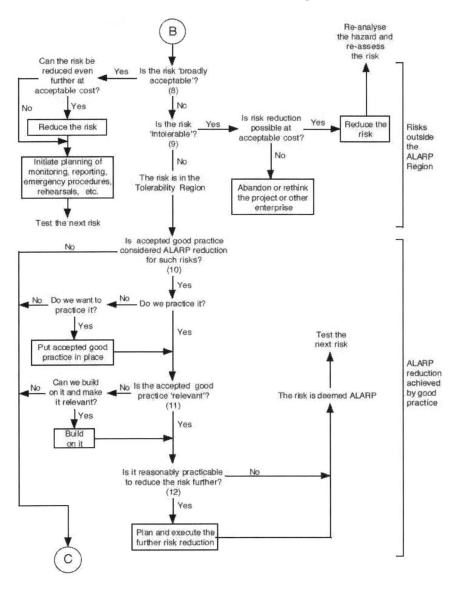

Fig. 2. Decisions for Risks Outside the Tolerability Region or for which Good Practice Provides Reduction ALARP

Hazard analysis and risk assessment should be carried out at a number of points in the life cycle of a system, and some of these will include ALARP decisions. The goal is for all risks to be ALARP when a system is deployed and subsequently during deployment, and it is the responsibility of the system's stakeholders to ensure this. Which stakeholder carries primary responsibility depends on circumstances; further, the distribution of responsibilities differs between stakeholders' organisations. Thus, this article addresses what needs to be done in making ALARP decisions, but not exactly who should make them or when they should be made; it uses the general term 'duty holder'.

The sections of Figure 2 designated by the large square brackets on the right are addressed in corresponding sections in the text below, and the numbers in round brackets (which continue sequentially from those in the flowchart of the previous article) are indexed by their equivalents in the subsection headings.

Some General Requirements

The ALARP Principle requires duty holders not only to reduce risks appropriately but also to judge what reduction is appropriate. Making such judgments demands knowledge and competence, and duty holders are expected to possess them. A further requirement is for organisational, and not merely individual, safety consciousness and behaviour. Senior management must set appropriate safety policy, define and nurture a proper safety culture, promulgate appropriate standards and practices, and manage the selection and training of competent, safety-conscious workers. In all cases, these are fundamental requirements, and their absence necessarily raises questions about the credibility of an organization's ALARP claims.

Risks Outside the Tolerability Region

Although the ALARP Principle covers the entire ALARP model (see Figure 1), the term 'ALARP decision' is often applied to the consideration of risks whose estimated values lie within the model's Tolerability Region - though such comparisons depend on the calibration of the ALARP model, as discussed previously (Redmill 2008). As well as addressing risks in the Tolerability Region, decisions must be made about those lying outside of the two thresholds.

Broadly Acceptable risks (8)

Even risks already in the 'broadly acceptable' region may be reduced further, and they should be if reduction can be achieved at reasonable cost. For example, the risk of loss of life due to fire in a building may be deemed broadly acceptable if the walls are of concrete to prevent a fire starting and doors are fire-proofed to stop it spreading. But further reductions could be achieved by introducing a smoke detection system, placing a fire extinguisher in every room, putting evacuation notices on the walls, including the rescue service's telephone number on the notices, conducting evacuation rehearsals, and more. Should these additional risk-reduction measures be taken? Almost certainly some or all of them should if their additional costs are small - and particularly if the building is open to the public. They should not be discarded as unnecessary just because the risk is already deemed to be low, but should be assessed for their value and cost.

Once it is determined that it is not reasonably practicable to reduce a broadly acceptable risk further, plans should be put in place to monitor it to detect any increase over time, due to change in either the system or its environment.

Intolerable risks (9)

A system must not be deployed if it is deemed to carry an intolerable risk. If the risk cannot be reduced, or if it cannot be reduced at a cost acceptable to the duty holder, either the objectives of the project must be altered so as to avoid the hazard throwing up the intolerable risk, or the enterprise must be abandoned. Continuation is therefore not merely a safety matter but also a commercial one; the question arises: is it worth reducing the risk sufficiently to save the project?

If the risk is reduced into the ALARP region, the new residual risk must be resubmitted to full analysis, starting with hazard identification - because it is not unusual for risk mitigation activity to throw up new hazards.

Good Practice

The HSE reports that in most situations ALARP decisions 'involve a comparison between the control measures a duty-holder has in place or is proposing and the measures we would normally expect to see in such circumstances, i.e. relevant good practice' [HSE 2008a]. Thus, great reliance is placed on good practice, which the HSE [2008d] defines as 'the generic term for those standards for controlling risk which have been judged and recognised by HSE as satisfying the law when applied to a particular relevant case in an appropriate manner.' In short, it is inefficient and costly to re-invent risk-reduction measures for the same equipment or activities, and it makes sense to employ measures that have already been proven.

Good practice may be documented, e.g. as HSE guidance, an approved code of practice, or a standard, or it may be unwritten and still recognised, e.g. if it is 'well-defined and established standard practice adopted by an industrial/occupational sector' [HSE 2008d].

The HSE refrains from requiring the use of 'best' practice for various reasons. For example, because there is seldom agreement on what practice is best; because what may be best in one circumstance may not be best in a slightly changed setting; and because a code of practice that is new but better than all others is unlikely to be widely used, standardised, or even widely known. However, a conscientious duty holder will strive to attain best practice, whatever it is perceived to be.

According to the HSE, the use of accepted good practice is, in most cases, a sufficient basis for a claim that a risk is ALARP, so it makes good sense for a duty holder to explore the appropriateness of good practice. The question is therefore asked, at point (10) in Figure 2, whether accepted good practice exists for the management of the sort of risk that is under consideration. Then, as it is crucial that the good practice is actually practiced by the duty holder, the further question of whether the stakeholder's organisation uses it (or will now use it) is asked. The answer 'no' to either question precludes an ALARP claim on the basis of good practice.

But even when the answer to both questions is 'yes', the question of relevance arises - at point (11) in Figure 2. Practice only becomes accepted as good after proven success in use with particular equipment in defined circumstances. The amalgam of rules, tasks and competences that compose accepted good practice in one circumstance would almost certainly require change if either the equipment in use or its operational environment were different. New hazards could be presented by a new equipment model that works similarly to its predecessor, or by existing equipment operated in a changed environment or put to a new use. So any such changes require study of whether the practice remains valid or whether it too requires amendment.

For the same reason, the proper use of standards, procedures, and other documented recommendations requires knowledgeable governance [Redmill 2000]. Every application - and every changed application - demands specific tailoring and guidance.

Thus, accepted good practice must be demonstrated to be relevant to the present circumstances, where 'relevant' is defined by the HSE [2008c] as 'appropriate to the activity and the associated risks, and up to date.' If it is not relevant, it may, in some cases, be tailored to become so; otherwise other risk reduction measures must be employed. Further, even if accepted good practice is employed and relevant, the decision that it reduces the risk under consideration ALARP should not be automatic; it requires consideration, acceptable evidence, and justifiable argument. It may require consultation with experts or approval from a higher authority.

Finally, no risk should be deemed ALARP unless the possibility of further reduction has been explored, as indicated by the question at point (12) in Figure 2. Once good practice has been followed, consideration should be given to whether more could be done to reduce the risk. If there is more, the presumption is that duty-holders will implement these further measures, but the reasonable practicability of doing so requires the application of first principles to compare the risk with the sacrifice involved in further reducing it.

Beyond Good Practice

There are numerous risks, across all industry sectors, for which good practice is inapplicable - for example, in the case of a new technology, or when elements of existing technologies or systems are interfaced. Then ALARP decisions must be made by comparing options for risk reduction. The main steps in this process are shown in Figure 3 (which is a continuation of Figure 2), with the following supplementary notes expanding on them.

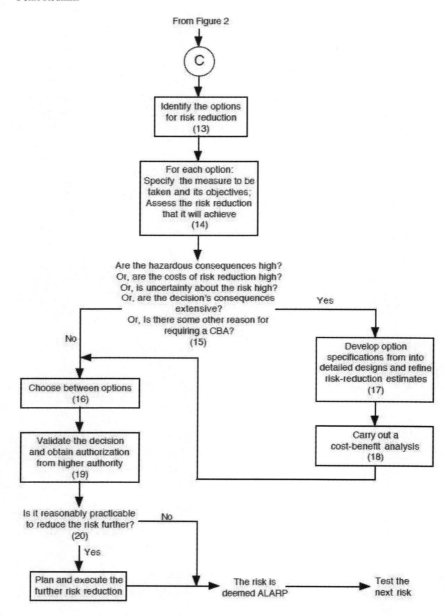

Fig. 3. Decisions on Risks for which Good Practice Does Not Provide Reduction ALARP Identifying and specifying options (13, 14)

The first step is to identify all feasible options. This requires good knowledge of the system under consideration and understanding of the mechanism of the risk in question. Without such knowledge and understanding, the credibility of decisions may retrospectively be questioned, particularly if an accident has occurred, so it is important for all whose expert judgment is relied on to possess adequate and relevant competence. Indeed, this should be documented (in the safety case, if there is one) so that decisions will later be seen to be credible.

Then, in order for comparisons to be made, a clear specification should be produced for each option, defining what measures would be taken (e.g. a guard will be installed on the machine) and what they are expected to achieve (e.g. the guard will prevent damage to operators' hands), and an assessment carried out to determine what risk reduction is anticipated (e.g. the frequency of damage to hands will be reduced from N per year to M per year).

Mode of decision between options (15)

Neither UK law, which requires risks to be reduced 'so far as is reasonably practical' (SFAIRP), nor the ALARP Principle, provides definitive guidance on when the cost of reducing a risk is disproportionate - or grossly disproportionate - to the benefits derived from reducing it. In the end, the decision must depend on expert judgment. In most cases, the circumstances of the risk are understood, the costs and consequences are not excessively high, and the effects of the decision do not extend outside the company. Then a first-principles approach, supported by knowledge and experience, is likely to lead to an ALARP decision that is justified by the evidence and logical argument.

But in some cases, for example in complex situations, in high-hazard industries and new technologies, and when the effects of the decision would be extensive, decision-making is less straightforward and comparisons between options require more detailed evidence. Then cost-benefit analysis (CBA) is used to provide additional information - not on the technicalities of risk reduction, but to support decisions on economic proportionality. Point (15) on Figure 3 is where it is decided whether or not CBA is required.

ALARP decisions by professional judgment (16)

In seeking to reduce industrial risks, there is a recognized order of preferences in the choice of measures to be taken:

- Avoid the hazard altogether, for example by replacing a toxic chemical with a non-toxic one;
- Address the hazard at its source by creating an inherently safe design, for example by introducing an interlock to prevent the hazard from giving rise to an accident;

- Introduce engineering controls, such as by collecting dangerous waste products and separately managing their safe disposal;
- Apply collective protection measures, for example regulatory controls such as exposure limits - and in any case, any applicable regulatory rules should be applied and adhered to;
- Provide specific protection for individuals exposed to the risk, for example special clothing and equipment;
- Impose protective procedures, such as signing-in on entry.

For a given risk, a measure of the first type should be chosen if one exists, unless its cost is grossly disproportionate to its benefits. If none exists, or it is grossly disproportionate, a measure of the second type should be selected unless one does not exist or it is grossly disproportionate. And so on - though the application of a higher-level measure does not preclude the additional application of a lower-level one. Thus, the onus is on the duty holder not simply to balance the costs of each possible measure against its benefits, but to implement the safest measure unless it involves a grossly disproportionate sacrifice - in which case it may be deemed not reasonably practicable.

When the circumstances, as at point (15) on Figure 3, are appropriate, an expert, using this hierarchy of preferences as guidance, could in most cases arrive at a sensible and justifiable ALARP decision. It is important, then, to document the argument and evidence for the decision, to seek validation of it by another expert, to obtain authorization from an appropriate level of seniority - as at point (19) - and to document all this information in the safety case.

Finally, as observed earlier, no risk should be deemed ALARP unless the possibility of further reduction has been explored, as indicated at point (20) in Figure 3.

Cost-benefit analysis (17, 18)

When the answers to the question at point (15) in Figure 3 are 'yes', the HSE [2001] recommends employing a cost-benefit analysis (CBA). Then, in order to facilitate derivation of the best possible estimates of costs and benefits, and so increase confidence in comparisons, the specifications prepared at point (14) must be developed into detailed designs and the estimates of expected risk reductions must be further refined.

The costs to be aggregated in a comparison are all those that are 'necessary and sufficient to implement the measure' [HSE 2001]. Included are the costs of implementation of the measure, operation and maintenance, as well as any losses in time or productivity that are directly due to the measure. From them should be subtracted any gains resulting from the measure, for example those arising from improved productivity and a decrease in accident costs. The costs of safety assessment and justification of the measure should not be included.

The Rail Safety and Standards Board has provided a table of advice on the applicability of costs and benefits [RSSB 2008].

Like costs, some benefits may be readily translatable into monetary terms - when they are based on marketable goods. But, in the field of safety, benefits may take the form of reductions in deaths or injuries, whose values are not determined from what they were sold for in the recent past. Yet, comparisons between costs and benefits depend on both being expressed in the same units. Thus, health and safety benefits must be translated into monetary terms. There is no universally agreed way of effecting this translation, and all methods are based on 'judgments', which are open to manipulation and often contentious. Clearly, whatever equation is used must be acceptable to all stakeholders, who consist, at least, of the duty holder, those at risk, and the appropriate regulator.

A method often employed is 'contingent valuation', in which representative samples of the population are asked what they would pay for an incremental increase in a defined form of safety, with values being derived from analysis of their answers. This is the approach taken by the UK's railway industry, which annually reviews its 'value of preventing a fatality' (VPF). The VPF is then applied to other safety benefits via the equivalence formula: 1 fatality is equivalent to 10 major injuries, 200 reportable minor injuries, and 1000 non-reportable minor injuries [RSSB 2008].

Since costs and benefits cover the whole of a system's life, accuracy of comparison demands that all be translated not only into monetary terms but also into present values. This requires discounting factors, whose choices are, necessarily, based on assumptions that may not reflect reality - or they may appear reasonable at the time but be undermined later by changes in economic circumstances. It should also be observed that results, and therefore comparisons, are sensitive to their values. Care should therefore be taken in their choice.

The results of a CBA are intended to be justifiable estimates of the costs and benefits of the identified options for risk reduction. However, given the several variables whose values are subjectively chosen and to which the results are sensitive, they cannot be considered definitive.

ALARP decisions informed by CBA

A decision informed by the results of a CBA cannot avoid expert judgment. Indeed, a CBA is not the sole decider, but only one source of information input to the decision-making process. The hierarchy of preferred options, introduced above, is still the primary basis of choice. CBA results provide evidence to inform and support judgments on the proportionality between the costs and the benefits of an option that has been chosen from the perspective of safety.

When uncertainty about the risk is high, a CBA should not be used to reject all options for risk reduction on the grounds of cost. Lack of proof, or evidence, of risk should not be taken to imply the absence of risk, and any concessions should be made on the side of safety. In other words, the precautionary principle should be applied.

Finally, as in choosing between options without CBA, the final choice of risk-reduction measure should be validated, authorization from a higher authority should be sought, and the possibility of further reasonably practicable risk reduction should be explored.

New hazards created

In making ALARP decisions, attention should be paid to the potential side effects of risk-reduction measures. It is a bonus if an effect is the reduction of another risk. But it may also happen that another risk is increased, in which case the costs of the further measures needed to reduce the second risk must be included in the costs of mitigating the first. Analysts must remain aware of the possibility of such adverse side effects by taking a holistic approach [HSE 2008c].

However, if a new hazard is created by a risk-reduction measure, the HSE [2001] advises that it should be treated separately - except that, if it is found not to be reasonably practical to reduce it ALARP, the first measure cannot be taken.

Summary

This article has explored the processes of making ALARP decisions and identified the requisite considerations. Using flow charts, it showed how risks outside the ALARP region of the ALARP model are handled. It discussed the importance and limitations of 'good practice'. It addressed the use of professional judgment and how this may be supplemented by cost-benefit analysis, which should be a contributor to decision-making but never the sole decider. The article calls on published advice and puts it into the context of good engineering practice.

Felix Redmill provides training in safety engineering and management principles, risk, and project management.

References

HSE [2001]. Health and Safety Executive: Reducing Risks, Protecting People. HSE Books.

HSE [2008a]. Health and Safety Executive: ALARP At A Glance. HSE Leaflet (available in pdf from http://www.hse.gov.uk/ - updated 08.04.08)

HSE [2008b]. Principles and Guidelines to Assist HSE in its Judgements that Duty-holders have Reduced Risk As Low As Reasonably Practicable. Health and Safety Executive leaflet (available in pdf from http://www.hse.gov.uk/) (updated 08.04.08)

HSE [2008c]. Policy and Guidance on Reducing Risks as Low as Reasonably Practicable in Design. Health and Safety Executive leaflet (available in pdf from http://www.hse.gov.uk/) (updated 08.04.08)

HSE [2008d]. Assessing Compliance with the Law in Individual Cases and the Use of Good Practice. Health and Safety Executive leaflet (available in pdf from http://www.hse.gov.uk/) (updated 08.04.08)

Redmill, Felix [2000]. Installing IEC 61508 and Supporting its Users - Nine Necessities. Fifth Australian Workshop on Safety Critical Systems and Software, Melbourne, Australia

Redmill, Felix [2008]. Assessing the Tolerability of Risk - Overview of the ALARP Principle. Safety Systems, Vol. 18, No. 1

Redmill, Felix [2009a]. The ALARP Principle: Legal and Historical Background. Safety Systems, Vol. 18, No. 2

Redmill, Felix [2009b]. Essential Preliminaries to ALARP Testing. Safety Systems, Vol. 18, No. 3

Efficiency and Effectiveness of the ALARP Principle

Felix Redmill

This is the third of three articles by Felix Redmill on the subject of ALARP. All three articles are presented together and counted as one in this volume.

First published in Safety Systems 20-1, September 2010

Review

The principles of modern safety thinking that underpin the ALARP Principle - such as a risk-based approach, self-regulation, and the need to achieve and demonstrate appropriate safety in advance - began their evolution within the English Common Law, were made explicit by Lord Robens in his review of the regulation and management of occupational risks [Robens 1972], and were then enshrined in statute in the Health and Safety at Work, Etc. Act [HSW Act 1974], which requires risks imposed on others to be reduced 'so far as is reasonably practicable' (SFAIRP). A law merely states what must be done; it is left to those who would abide by the law to determine how to do it and to other authorities (assessors, regulators, and the courts) to determine whether it has been done in conformity with the law. In the absence of other guidance on how to go about reducing risks SFAIRP, the UK's principal safety regulator, the Health and Safety Executive (HSE), proposed the ALARP Principle, which is a strategic approach based on a model.

The series in this Newsletter, of which this is the seventh and concluding article, commenced in Vol. 18, No. 1, with an overview of the ALARP Principle. This was followed by the Principle's historical and legal origins, and then its application in all types of risk-tolerability situations were analysed. Finally, a number of topics that necessarily interact with the ALARP Principle, and on which the Principle depends, were discussed, and it was shown that each of

© Felix Redmill 2010.
Published by the Safety-Critical Systems Club. All Rights Reserved

these topics and their interactions with the ALARP Principle offer opportunities for further research.

Following this exploration of the ALARP Principle, its nature and purpose, its background, and its application, there are two important questions to be asked: how efficient is it, and how effective is it?

Efficiency of the Principle

Though easy to understand, the ALARP Principle is not so easy to apply. It offers a straightforward model to facilitate risk-tolerability decisions, but its use requires a great deal of professional knowledge and judgment and depends on the handling of a number of risk-related factors.

The decision-making processes were presented in the series of articles as three flowcharts. As these are intended to cover all situations, they contain a large number of decisions points. However, in engineering safety, uncertainty is mostly low and the determination of appropriate ALARP decisions follows readily from the proper identification and sensible analysis of available options. Indeed, according to the HSE [2008], most ALARP decisions are based on the use of accepted good practice.

But when uncertainty is not low, decisions are not so easily arrived at, and confidence in them may not be high. To derive them, not only is a sound knowledge of risk and its analysis required, but so also is knowledge of the factors to be considered in applying the ALARP Principle, many of which are not within engineering or risk curriculums and not within the compass of most engineers. Some (e.g. reasonableness, practicability and gross disproportion) are legal; some (e.g. cost-benefit analysis) are economic; some (e.g. societal concerns) are sociological; and some (e.g. decision making, and the effects of human biases on risk analysis) are psychological. Thus, in order confidently to arrive at just, socially acceptable, legal ALARP decisions, there is a requirement not only for knowledge in the various subjects but also for an open, observant and analytical approach to bring them together in a manner appropriate to the situation in hand.

Then there are questions of confidence - in the results both of risk analysis and decisions taken - and how to determine it and how to express it.

Applying the ALARP Principle when uncertainty is high is a non-trivial exercise. Not only are risk and its analysis tricky subjects, but, as already pointed out, engineers and other risk analysts and managers may possess no more than a superficial knowledge in the domains of the related factors.

Effectiveness: ALARP and SFAIRP

But what of the ALARP Principle's effectiveness? What confidence can there be that a risk deemed ALARP would also be judged to have been reduced

SFAIRP? Can the two concepts be said to be identical? They cannot. As pointed out earlier in the series, they were defined by different parties (the lawmakers and the safety regulator) for different purposes (stating a legal requirement and offering guidance on a strategic approach to meeting it). But does one imply the other? No. Meeting either is never an incontrovertible fact, proven by some irrefutable logic; rather, it is a matter of judgment. It may reasonably be assumed that the party (an individual or a team) that judges a risk to be ALARP would, a moment later, judge the same risk to have been reduced SFAIRP. But there are several factors why a risk judged ALARP now may not, later, be judged to have been reduced SFAIRP.

First, there is the matter of time. The law is likely to be invoked after an accident has occurred, and this would be some time after a system has been deployed (with its risks deemed ALARP). ALARP is now; SFAIRP judgements are made later, in a court of law, to settle a civil dispute or a criminal prosecution.

Second, the two judgments are made by different people. The ALARP judgment is likely to be made by practitioners, assessed by third-party independent safety assessors and, perhaps, approved by a regulator, whereas a legal SFAIRP judgment is made by a jury or judge.

Third, there is the matter of information. The legal judgment is made in retrospect, based not only on what informed the ALARP decision but also on what has come to light since and on opinions offered by expert witnesses. Different questions asked by barristers and different assertions made by expert witnesses could make all the difference to a judgement by a jury or judge. The scrutiny exercised in a court of law may also lead risk creators to recognise that there was information that they should have collected or considered but didn't and actions that they should have taken but neglected to take.

Fourth, there is the matter of perspective. Practitioners making ALARP decisions are likely to start with the desire to demonstrate that their system is adequately safe - i.e. that its risks are all tolerably low - whereas a jury is likely to look for evidence that it was not. The difference in perspective can be significant.

Fifth, there is the matter of subjectivity. Whatever the strength of other factors, ALARP and SFAIRP decisions depend on judgment and therefore involve subjectivity. A practitioner may place greater or less emphasis on this or that piece of information, or on the opinion of this or that advisor, while a jury or judge may be swayed by the persuasiveness of the advocate for the prosecution or the defence, by the logic of this or that expert witness, or by the framing of this or that piece of evidence. The greater the uncertainty surrounding a risk, the greater is the influence of subjectivity in a tolerability decision. Subjectivity is an essential ingredient in professional judgment, but in court, later, a jury may subjectively conclude that the decision-maker should have inclined in a different direction.

Sixth, there is the matter of context. Engineers and other duty holders must make risk decisions that meet legal requirements; juries and judges must deliver determinations of whether or not those decisions do, or did, in fact satisfy the law. Practitioners make their decisions in a technical context and juries and judges make theirs in a legal context. As mentioned earlier, the ALARP Principle is intended as a bridge between the legal and the engineering to assist the practitioner to interpret the legal requirements, and then as a bridge back to the legal in making the claim that the requirements have been fulfilled. In the two crossings, there is plenty of scope for misunderstanding. A record of earlier ALARP judgment may be evidentially useful, but it cannot guarantee a favourable court finding later.

The ALARP Principle is not a formula for the provision of consistent or repeatable results. Nor is the law. Both are designed to cover all possible situations and, therefore, both require judgment whenever they are employed. They both depend on determinations of practicability and reasonableness, which, as explained above, differ according to the case in hand and, in all cases, are subject to judgment. There can be no guarantee that the same ALARP decision would be arrived at by two different practitioners, and certainly none that an ALARP decision arrived at now in an industrial context would, later be judged by non-engineers in a legal context to have met the SFAIRP test. Indeed, whereas the foregoing discussion contrasts ALARP and SFAIRP decisions, it could, as easily and without change, be used to contrast ALARP decisions made by different people at different times.

The Importance of Confidence

It would seem that, in a trial on the legal, SFAIRP, side of the ALARP bridge, a good lawyer is worth more than a good engineer. But what must the engineer provide in order to give a good defence lawyer an advantage over the prosecution's good lawyer? The answer is: evidence that all was done that could reasonably have been done in the circumstances. Thus, there needs to be a record, preferably in a safety case, of what lies behind every ALARP decision, i.e. sound evidence of due diligence: not only the details of risk analysis and risk-management actions, but also the assumptions made and reasons for their validity, and an assessment of the evidence underpinning each decision and the justified level of confidence in it.

In particular, it is important to provide reasons not only for action but also for doing nothing. When an accident has led to an inquiry or a prosecution, inaction may, in retrospect, give the impression of negligence, and negligence may be hard to refute. There needs to be evidence, first that it was not negligence and, second, that inaction was justified. More than that, the engineer should be able to claim to have been 'as confident as reasonably practicable' (ACARP) in each ALARP decision. To increase the likelihood of favourable

SFAIRP judgments, ALARP decisions should be supported by ACARP arguments.

Further Work and Final Words

The ALARP Principle is a tool for the assessment of risk tolerability, and its operation depends on the manner in which its users define and address other aspects of risk and its analysis and management (the more important of such topics were discussed in Redmill [2010a, 2010b]. Such definitions and manners of address create assumptions that become implicit in each application of the ALARP Principle. Further, they vary between practitioners for a number of reasons, including a lack of knowledge of risk principles and of other relevant factors. Thus, improved use of the ALARP Principle - and, indeed, improvement in the assessment of risk tolerability in general, whether or not the ALARP Principle is employed as a tool - calls for better education in the subject of risk, better knowledge of the related topics, and research into some of those topics.

Education would be improved by teaching risk to all science and engineering undergraduate and post-graduate students (as well as others, such as those studying medicine, law and psychology). Such teaching should provide adequate instruction in the theory; it should also emphasise that techniques are not the principal components of the subject but tools to be chosen appropriately and only used with understanding.

Finally, it should be mentioned that the ALARP Principle is not essential or mandatory. It is quite possible to carry out risk tolerability assessment and to satisfy the law without employing it. Indeed, many do so, for many who must reduce their risks have never heard of it. It is the principle of taking a risk-based approach and being able to demonstrate that all risks have been reduced SFAIRP that are fundamental, not the use of a particular technique or model. Having said that, it is reasonable to assume that, as the ALARP Principle is the HSE's model, a claim to have satisfied it would carry weight in a court of law.

Felix Redmill provides training in safety engineering and management principles, risk, and project management.

References

HSE [2008] Health and Safety Executive: ALARP At A Glance. HSE Leaflet (available in pdf from http://www.hse.gov.uk/) (updated 08.04.08)
HSW Act [1974] The Health and Safety at Work Etc. Act. HMSO, London
Redmill, Felix [2010a] Discussion of Some ALARP-related Topics. Safety Systems, Vol. 19, No. 2, January
Redmill, Felix [2010b] Discussion of More ALARP-related Topics. Safety Systems, Vol. 19, No. 3, May
Robens, Lord [1972] Safety and Health at Work: Report of the Committee, Cmnd 5034. HMSO, London

ALARP Explored

A full paper, consisting of the seven articles (suitably edited together) published in this newsletter in consecutive issues since September 2008, has now been published as a Technical Report at the University of Newcastle upon Tyne. Titled ALARP Explored, its reference mumber is CS-TR-1197, and it may be downloaded at http://scsc.org.uk/p113.

Armageddon and Other Failure Modes

John Ridgway

First published in Safety Systems 22-2, January 2013

When it comes to doomsday scenarios and visions of technological dystopia, we all seem to have febrile imaginations, fed by a diet of science fiction and marshalled by a healthy contempt for mankind's hubris. It is no wonder, therefore, that the proposed commissioning of CERN's Large Hadron Collider (LHC) was accompanied by a seemingly obligatory outcry, warning of a demise that would make Armageddon look like an easy day out for a lady. Undoubtedly, worlds would crumble, as the shrieks of 'I told you so' wafted out to infinity and beyond.

I have to confess that my reaction to the dire warnings of black holes and sundry cataclysms was to treat the whole thing as a big joke. I tutted scornfully at the ill-informed hysteria and I wearied that the good folk of CERN should be so pilloried by the crank and file of modern society. So I decided to look into the subject further, mainly so I could bamboozle and humiliate the next doomsayer who dared to repeat such ridiculous claims in my presence. However, as I delved deeper, a picture emerged that left me somewhat chastened. The true story behind the safety of particle accelerators is firmly grounded in the real world of engineering and, as it turns out, even the more far-fetched concerns had a respectable provenance. Furthermore, the doomsday speculations left me wondering whether calculations of risk are always sensible.

In this article I can only offer you a brief overview of the subject and, as always, I have to declare my layman status. Nevertheless, once I have finished warning you of the hazards of sticking your head into a particle beam, and I have revealed a recipe for the destruction of the Universe, I wonder if you will share my sense of humility.

Beware the Beam

Any system that requires the control of energy may be seen as safety-related, simply because control failure, or failures in protection, can result in the energy's harmful dissipation. When thinking about particle accelerators, the classic example would be a scenario in which the particle beam intersects human flesh.

© John Ridgeway 2013.
Published by the Safety-Critical Systems Club. All Rights Reserved

Given the design of modern accelerators, you would think that such an event would be impossible; and if it did happen, the results would surely be fatal. But you would be wrong on both counts.

In 1978 a researcher named Anatoli Bugorski was working on a malfunctioning component of the U-70 synchrotron at the Soviet Institute for High Energy Physics. Whilst leaning over the piece of equipment, he stuck his head in the part through which the proton beam was running. Normally, a radiation dosage of 500-600 rads would be fatal. However, protons are relatively heavy and do not dissipate so easily when entering the body (hence their use in radiotherapy). As a result, a dosage of 300,000 rads resulted in no more than paralysis of the left side of his face, deafness in the left ear, and long-term problems with epilepsy. Anatoli thinks himself lucky. Nevertheless, he might reflect upon the fact that the accident was eminently avoidable. It resulted from a failed safety control, and any half-decent HAZOP and risk analysis could have foreseen such a mishap.

Radiation Risks

Of course, the radiation risks resulting from particle acceleration go far beyond those that may be the undoing of an inquisitive engineer. All charged particles emit radiation under acceleration and so energy loss is a fact of life for experimental high energy physics. As a result, operational efficiency is a significant engineering challenge, as are the health hazards resulting from both intentional and unintentional radiational loss. Such radiation hazards fall into two categories:

Prompt Radiation - Generated by intentional and accidental beam losses in beam line components. This can be very intense but is present only when the accelerator is on. Radiation scattered through air (so-called 'skyshine') may affect areas which are not in direct sight of the accelerator and beyond the boundary of the accelerator site.

Induced Radioactivity - Generated mostly in highly irradiated beam-line components, coolant systems, shielding materials, ground, groundwater and ambient dust/air.

The management of these hazards falls squarely within the province of standard health physics. Consequently, we see the use of barriers and gates designed to keep people away from radiation (though it has to be said that electrical hazards constitute by far the most significant risks facing operational staff). We also see extensive use of beam-containment interlocks, shielding, radiation monitoring devices and special waste management. More to the point, the physics involved is well-understood and standard techniques for hazard identification and analysis and risk assessment should be more than adequate.

Mechanical Failures

Still within the realms of the mundane, we need to consider what physical damage might be done as a result of mechanical failures occurring during operation, and whether such damage could result in harmful events. Sadly, within this category, the LHC provides an excellent and somewhat expensive example.

On 19 September 2008, within only a few days of the LHC being brought into operation, a fault in an electrical connection between two beam controlling magnets caused an abnormal termination of magnet operation such that part of the superconducting coils entered the normal (i.e. resistive) state. The event known as a magnet quench resulted in a rapid release of energy and an explosive boil-off of the surrounding cryogenic fluids (liquid helium); these then vented into the beam tunnel. The resultant damage was extensive (over fifty magnets in total were damaged) and the operational delays incurred, together with costs of repair, were unwelcome to say the least. Thankfully, however, no one was harmed; leaks of liquid helium rapidly vaporize and could have frozen or choked anyone present.

Again, the point to be made here is that good old-fashioned risk-based management applies. It's not as if magnet quenches were previously unheard of.

Doomsday Risks

Strangely, on the day that the LHC superconducting magnets were losing their cool, the international press were still more concerned by the menace of mini black holes marauding the Swiss countryside. So where did this preoccupation come from?

In fact, the spectre of doom was first raised not for CERN's LHC but for one of its predecessors - the Brookhaven Relativistic Heavy Ion Collider (RHIC) in the USA. As a consequence of these concerns, a safety committee was set up by Prof. John Marburger, Director of Brookhaven National Laboratory, which reported in September 1999. When similar concerns were echoed for the LHC, a further investigation was instigated by CERN, which built upon the RHIC investigation. An initial report was published by CERN in 2003 [1], and a follow-up, extended report appeared in 2008 [2]. The 2008 report was categorical in its findings:

'We conclude by reiterating the conclusion of the LHC Safety Group in 2003: there is no basis for any conceivable threat from the LHC. Indeed, theoretical and experimental developments since 2003 have reinforced this conclusion.'

In coming to this conclusion, the LHC Safety Group had considered the creation of four hypothetical objects, any of which could seriously ruin your day:

Microscopic Black Holes

According to conventional General Relativity theory, it is not possible to create a black hole as a result of a collision between two protons; the gravitational forces involved are just not great enough. It is only when one considers speculative gravitational theories based upon more than four space-time dimensions that such a possibility cannot be discounted. Even so, the LHC Safety Group calculated that the energy generated by two protons colliding within the LHC is equivalent to two colliding mosquitoes, and so any black hole that might be created by the LHC would have to be much smaller than any known to astrophysics. Nevertheless, a black hole is a black hole and you wouldn't want one in your back yard.

Thankfully, however, all black holes are destined eventually to evaporate to nothing as a result of Hawking radiation. The key point here is that the smaller the black hole, the faster the rate of decay, such that the sort of black hole that might be produced by the LHC would lose mass faster than it could accrete it. As a result, the black hole would decay to nothing long before it had the opportunity to escape the walls of the collider. The LHC Safety Group also backed up this theoretical argument by pointing out that ultra-high energy proton collisions are commonplace within the cosmos and so the LHC experiment has, in effect, already been repeated 1031 times since the creation of the Universe. If the creation of destabilising mini black holes was so easy, then why do we see no astronomical evidence for them?

Negatively Charged Strangelets

The LHC research programme includes the collision of heavy ions, the purpose being to reproduce the plasma formed at the birth of the Universe. Most of the plasma's nuclei will be constructed from quarks that constitute 'ordinary' matter, i.e. from the up and down quarks. However, other, more exotic, quarks will also be produced, including the so-called strange quark. Nuclei that combine both up/down and strange quarks are known as strangelets. Strangelets have already been created experimentally (for example, by the RHIC) but have always proven to be unstable, decaying in about a nanosecond. The problem is this: what if stable strangelets were to be produced? Such material might be more stable than the ordinary material with which it were to come into contact. A runaway conversion to strange material would result, and the energy released during the conversion process would be enough to turn the Earth into a molten lump.

The technical argument offered in the CERN 2008 report is somewhat long and complicated, but it amounts to this: Firstly, no stable strangelets have ever been created in the laboratory, despite several years of RHIC operation. Secondly, even if stable strangelets were possible, theory strongly suggests that they must be positively charged. Such strangelets would then be harmless as they would repel, rather than attract, the ordinary nuclei around them. Finally, a similar argument to that used for microscopic black holes is invoked: stable, negatively charged strangelets cannot be that easy to produce, otherwise we would surely see evidence for their existence in nature. For example, the Moon has existed for billions of years under the bombardment of high energy cosmic material and yet has remained resolutely non-strange.

Magnetic Monopoles

Of all the exotic objects posited by the CERN study, magnetic monopoles are easily the weirdest. I do not have space to do them justice here, but all you really need to know about them is that they are a purely theoretical invention and, if they did exist, they would be far too massive to be created in any existing particle accelerator, nor indeed any conceivable future accelerator. Furthermore, if they could be produced, their Earth-bound orgy of destruction would be short-lived since the energy generated by the destruction would propel the beastie out into space long before you could say, 'Oh dear, what can that matter be?'

Vacuum Bubbles

Those of you who have a general interest in quantum mechanics will already be aware of the concept of the vacuum ground state. Put simply, a physicist does not consider empty space to be empty but permeated instead by fields mediated by virtual particles that are constantly flitting in and out of existence. The vacuum, therefore, has an energy state. The presupposition is that the vacuum energy that universally appertains is at ground state since, if it weren't, the state would be unstable and would have decayed before now. But what if the vacuum energy state is currently a false minimum, i.e. a meta-stable state? And what if a high energy event in a particle accelerator is all we need to destabilise the existing vacuum to create a bubble of 'true' vacuum. The bubble would then expand at approaching the speed of light until it eventually engulfed the observable Universe. The problem is that such a new order could not hope to support physics and chemistry as we know them, let alone biology. A more absolute catastrophe is impossible to conceive. The only crumb of comfort is that if it hasn't happened already in the face of cosmic turmoil, then there is little to fear from the relatively puny efforts of the LHC.

So When is a Risk Not a Risk?

For most of the hazards posed by particle accelerators one can see how standard, risk-based engineering would apply. However, doomsday hazards seem to me to be of a fundamentally different type. If risk is a function of probability and impact, then just how low must probability be for the risk of destruction of the Universe (or even the Earth) to be deemed tolerable? It strikes me that the only acceptable level is zero, which is to say, there is no risk. Rather than trying to quantify the risk and reducing it ALARP, one has to convince everyone that the risk is not even an issue. And that is very hard to do once one has entertained the possibilities. The CERN Safety Group reports make a convincing case but, somehow, I am less reassured to have seen that the doomsday scenarios were taken seriously at all.

So we are then left more concerned with measuring our confidence in the scientists, and the problem is that the scientists employed, although pre-eminent, tend to be somewhat equivocal in the presentation of their reports. On the one hand, within the executive summaries, they dismiss the risks categorically, because that is both realistic and the politically important thing to do. On the other hand, aspects of their arguments are couched in cautious language, because that is what their scientific training and instincts dictate. No argument is devoid of assumptions and those made by the scientists appear reasonable enough, but they do seem to leave the tiniest room for doubt. So their message seems to be "Destruction is absolutely impossible - as far as we know".

Personally, I don't trust ultra-low probabilities when combined with epistemic uncertainty. The CERN report says there is no conceivable threat, but what it really means is that the conceived threats were all unrealistic. Under these circumstances I am almost tempted to be risk averse and say that if we only avoid the destruction of one universe, then the effort would probably still be worth it.

References

[1] Study of Potentially Dangerous Events During Heavy-Ion Collisions at the LHC: Report of the LHC Safety Study Group, Blaizot et al, CERN 2003-01
[2] Review of the Safety of LHC Collisions, Ellis et al, CERN-PH-TH/2008-136

John Ridgway is a retired analyst who, from time to time, dabbled in matters of software development, quality assurance, security management and systems safety engineering. To the best of his recollection, he has never worked on a particle accelerator.

Software testing and IEC 61508 – a case study

Ian Gilchrist

    IPL
    Bath, UK

First published in Safety Systems 18-1, September 2008

This article describes the testing activities carried out by IPL in the implementation of a system to the safety standards mandated by IEC 61508. The bare facts of the project are that the system has been evaluated as SIL (safety integrity level) 1, and was coded in C++ with about 115 KLoC (thousand lines of code) produced, tested and delivered to the customer. The project started in late 2002, and there have been three phased deliveries, in January 2004, April 2005, and March 2007. References to IEC 61508-3 are to its first edition.

Highway systems and the NASS project

The client, the Highways Agency (HA), has responsibility for managing, maintaining and improving the motorway and trunk road network in England. To help with the task of avoiding congestion, several computer systems are already in place. The current system, used to control roadside equipment and monitor road conditions, is called NMCS2 (National Motorway Communication System 2). An ATM (Active Traffic Management) pilot project is currently being run by the HA on one section (in the West Midlands) of the UK motorway network. The aim of ATM is to make best use of existing road space to increase capacity and ease congestion by controlling traffic according to actual and predicted road conditions.

    In 2002 the HA went out to tender for the development of a new subsystem called the Network ATM Supervisory Subsystem (NASS), to form an additional element within the existing NMCS2 and future ATM systems. NASS takes real-time actual traffic flow data, combines this with historical traffic flow data, and then predicts future flows. If congestion is predicted, NASS will evaluate a number of predefined traffic control plans to avoid or minimise the predicted congestion, selecting the optimal plan. NASS will then issue requests for the settings of roadside signals and message signs to implement predefined traffic control plans.

The NASS contract was awarded to IPL in late 2002. The first milestone was a 'proof of concept' (PoC) system, which came in at about 20 KLoC and completed Factory Acceptance Tests (FAT) in December 2003. The second phase was for the production of a demonstrator system to be supplied to Traffic Engineering consultants working for the HA. The purpose of this system was to allow those consultants to refine the rules and algorithms internal to NASS for its safe and efficient functioning. This was delivered in April 2005, and was approximately three times the code size (60KLoC) of the initial PoC system.

The current (third) phase of NASS was delivered to the HA's West Midlands Regional Control Centre (RCC), located at Quinton, in March 2007. This is being gradually integrated with all the other systems running at the RCC with the aim, hopefully, of being able to demonstrate its worth by Summer 2008. At this point, NASS could be used to directly request sign and signal settings, thereby influencing drivers using the West Midlands motorway network with a view to reducing or mitigating congestion. Due to the target environment for NASS having evolved during its development, it is likely that closer integration with existing user interfaces will be necessary prior to making this step. Once this has been achieved, new variants of NASS could be produced and installed at other RCCs (of which there are seven) across the UK's motorway network. See Table 1 for a summary of the phases of the NASS project to date.

Phase	Name	Purpose	Delivered	Approx. code size
A	PoC	Prove NASS concept	Jan. 2004	20 KLoC
B	Demonstrator	Testbed for refinement of NASS rules	Apr. 2005	60 KLoC
C	NASS V1	For live use at West Midlands RCC	Mar. 2007	115 KLoC

Table 1. NASS project development phases

System safety and IEC 61508

The use of IEC 61508 was decided on by the HA as a result of consideration of the hazards involved. At the start of the NASS project, IPL engineers assessed that the safety integrity level appropriate for the project was SIL 1. This relatively low grading reflects the fact that NASS does not directly control any hazardous equipment, but is involved in issuing requests for traffic sign and signal settings – which can have safety consequences when acted upon.

NASS software design

IPL started work in early 2003. Having agreed the system requirements in detail for the PoC phase, software design followed a method based on use of UML. The design hierarchy led to the identification of sub-systems, which in turn comprise software components, which can be either executables or dynamic linked libraries (DLLs). Further OO (object oriented) design decomposition leads to the identification of classes, for which module specifications were created, ready for coding and testing by programmers. The module specifications detailed class public and private methods with code flow shown in pseudo-code, and class test plans.

The initial (PoC) phase design included 19 software components (six executables and 13 libraries) to make up the active NASS elements, comprising a total of 94 classes. The current phase delivery has grown to eight executables and 15 libraries, comprising a total of about 280 classes. The NASS system runs on Windows, and code production has been done using MSVC++ V6.

Testing strategy

The IEC 61508 standard calls [Clause 7.3.2] for validation planning, and furthermore [Clause 7.7.2.6] that 'testing shall be the main validation method for software'. Accordingly, the NASS project team together with the HA drew up a test strategy which included a formalised (i.e. under independent QA monitoring) approach to testing each and every entity at each and every identifiable stage of the project. Each entity has its own test plan, which details, as appropriate, the configurations, inputs and expected outputs which, when run successfully, will give the required confidence in the correct working of the entity under test. At the higher levels, the test plans were contained in a separate (version-managed) document; at the lower levels test plans were included in the design specification. Table 2 summarises the relationship between the principal design documents and the corresponding test specifications.

Design Document	Informs test plan for...
Existing NMCS Documentation	System Interaction Test
NASS System Requirements Spec (SRS)	Factory and Site Acceptance Tests
Architectural Design Spec (ADS)	System Integration Test
Sub-systems Design Spec (SSDS)	System Integration Test
Component Specification	Component Test (in the design spec)
C++ Class Specification	Module/Unit Test (in the design spec)

Table 2. Hierarchy of principal requirements and design documents

Throughout IEC 61508 [e.g. Clause 7.9.2.4] there are demands to the effect that test results should be generated to show that tests have been run and 'satisfactorily completed'. This put quite a necessity on the team to ensure that not only were they using tools that would make the various testing activities as easy and repeatable as possible, but also that the tests should, as far as possible, be self-documenting.

Code and module testing

Following classical V-model lifecycle principles and IEC 61508 [Clause 7.4.7, Software Module Testing] the first task for the IPL software engineers after programming the classes was to test them. For this purpose the IPL tool Cantata++ was used. This was partly because the engineers were familiar with it, and also because it gave all the functionality needed to test to the IEC 61508 SIL 1 standard. The basic requirement is to test every class in isolation and to demonstrate code coverage to the levels of 100% entry-point (every function/method called at least once) and statements. In fact, IPL took the reasonable decision to additionally test to 100% condition coverage. This is more work than the basic project SIL demanded but was felt by the developers to give a useful additional level of confidence.

Since every class had a number of external interfaces, not all of which could be stubbed, the Cantata++ 'wrapping' facility was vital in allowing the isolation testing to be completed as planned. Stubbing involves the replacement of an external function with a programmable simulation having the same interface. Wrapping allows for the 'real' external code to be included in the test build, but with the option to intervene at the interface in order to, for example, check values being passed out or alter values being returned to the code under test.

Integration testing

Cantata++ was also used for the next level up of testing, namely component testing. This formed the first level of integration testing, and was aimed at verifying the correct working of each NASS executable or DLL against specifications defined in the appropriate level of design. The testing involved calling public interfaces of the component under test, and stubbing or wrapping calls to external functions. Since NASS has its own database, the component tests included 'real' database code so that testing included the option to initialise the database and check that updates to it were as expected.

The project created, and has maintained, a fairly elaborate regression test facility, which has allowed for nightly builds and re-runs of each class and component test in the entire system. This has served well to enhance confidence in the change impact analysis system by ensuring that changes in one module are completely and properly compensated for in other affected modules and tests.

After testing the components the project test strategy called for sub-system testing. Since 100% coverage had already been achieved during unit testing, integration testing could be allowed to live up to its name – namely testing the integration of the software units. The team, with the agreement of the HA, determined that 100% entry point coverage was suitable to demonstrate the completeness of the integration tests. This is, in fact, exactly in-line with the IEC 61508 requirement [Clause 7.4.8.3] to demonstrate that, 'all software modules ... interact correctly to perform their intended function ...'

System-level testing

Following sub-system testing, the team carried out a further series of system integration tests, which were formally documented by IPL QA staff. These tests mainly served as a dry run to gain confidence before going into the customer-witnessed Factory Acceptance Tests (FAT). For the most recent delivery of the project, FATs took 20 days to run, much of which time was occupied by reconfiguring the system between successive test runs. It was noteworthy that the system integration tests ran smoothly and only revealed a few design anomalies.

The last layer of testing before the NASS was allowed to be installed at the RCC was called Interaction testing. This was carried out at the offices of another HA contractor, Peek Traffic, and involved running NASS on a rig which included NMCS2 equipment in exactly the same configuration as the live NMCS2 system. The intention here was to ensure that NASS could interact correctly with the other live systems at the RCC.

Lastly, Site Acceptance tests were held at the RCC to demonstrate that the live NASS was in fact operating correctly and safely with the rest of the NMCS2 system.

See Table 3 for a summary of testing levels within the NASS development project:

Name of tests	Testing	Tool	Comments
Module/Unit tests	C++ classes	Cantata++	283 in total. 100% Entry-point, statement and condition coverage. Test plan is in the class specification.
Component tests	Components (individual executables or DLLs)	Cantata++	23 of these in total. 100% Entry-point coverage. Test plan is in the component specification.
Integration tests	Aspects of ADS and SSDS	IPL-developed simulators	
System Testing	Entire System	IPL-developed simulators	Dry-run of Factory Acceptance Tests, witnessed by IPL QA staff.
Factory Acceptance Testing	Entire System	IPL-developed simulators plus HA 'Portable Standard'	Formal run of System Tests at IPL offices, witnessed by HA/consultants.
Interaction tests	Entire System	Test Rig at Peek Traffic	Formal run of Tests at Peek, witnessed by HA/consultants.
Site Acceptance Testing	Entire System	Live at RCC	Formal run of tests at RCC, witnessed by HA/consultants.
Regression tests	Re-runs of Unit and Component tests	Cantata++ and IPL-developed framework	Run nightly.

Table 3. Complete hierarchy of testing on NASS project, at third phase of the project

Conclusion

This article shows that a properly run IEC 61508 project, even at a relatively low SIL, can be demanding on test time. We wish to reinforce the point that testing is, on one level, about providing reassurance to developers that they can move with reasonable confidence from one stage of the project to another. At a different level, it can provide confidence to other stakeholders (e.g. the customer) that the system will work safely and reliably when installed on site. A good project will work from a test strategy (i.e. determine in advance at what stages and levels in the lifecycle testing should be carried out), and will demand that test plans exist for each entity to be tested; testing should not be ad hoc! Testing needs to generate results as evidence of test completion (i.e. be self-documenting). Furthermore, testing needs to be conducted in a repeatable fashion, because the one thing you can be sure of is the need to run and re-run tests at all levels many times.

IEC 61508 and Related Guidance - Uses and Abuses

David J Smith

Technis

First published in Safety Systems 21-3, May 2012

The Background to Functional Safety

For over fifteen years, the IEC 61508 guidance has spawned a raft of industry-specific documents disseminating much the same theme. The effect has been to outline the major aspects of targeting and assessing risk and to radically enhance the awareness of this branch of engineering. It is now almost unheard of for a major project not to include the identification of hazards and the subsequent risk engineering activities called for in the above guidance. This has led to risk targets (usually misleadingly referred to as 'SILs' - safety integrity levels) being placed on most of the elements of the supply chain, from the systems integrators down to the suppliers of equipment and instrumentation.

On the plus side there is now almost universal attention to safety integrity and a widening of assessment to encompass non-quantitative as well as quantitative factors. Cost is seen in better perspective due to the application of the ALARP principle and there is a wider availability of quantification tools.

On the negative side there is an obsession with the SIL 'word' without understanding its limited meaning as a metric for the application of non-quantitative assessment. There has been a dumbing-down of targeting methodology to enable all and sundry to 'have a go'. This seems strange when all other branches of engineering recognise the role of the specialist, to whom the calculations are entrusted. Not so with safety-integrity. Certification and the application of SIL targets is often taken to too low a level, such that 'bells and buzzers' are procured with integrity targets. Also, there is a frequent lack of focus in the choice of rigour, which leads to the 'paragraph by paragraph' mentality addressing each

© 2011 Technis.
Published by the Safety-Critical Systems Club. All Rights Reserved

and every statement in a standard with equal rigour, thus losing sight of priorities.

Desirable Outcomes

Universal attention to safety integrity

Prior to the development and publication of IEC 61508, quantified risk assessments were carried out in many industries. They involved the methods and tools described in the textbooks of that time and, in some cases, made use of in-house guidance and procedures.

The motivation for this work was to some extent voluntary and driven by the realisation, within an industry or organisation, that potentially hazardous events required analysis and prediction, leading to some visible attempt at mitigation.

Following the Flixborough (1974) and Seveso (1976) accidents, various aspects of legislation and guidance [1] provided additional impetus, culminating in the CIMAH (later COMAH) regulations in the UK. However, those regulations apply only to major industrial hazards and not the vast range of industrial product areas and applications. IEC 61508 (2000 and now 2010 version) has become so widely known that it is now rare for a product or process not to involve functional safety issues. Hazards are routinely identified, targets are set, and assessments carried out to establish if those targets are met.

Widening of assessments to encompass non-quantitative factors

Earlier assessments did not always involve establishing a target (i.e. risk of fatality) such that the assessment result could be deemed 'satisfactory' or otherwise.

Furthermore, assessments were largely quantitative. That is to say, they predicted the frequency of the event in question using available component failure-rate data. Whilst this might have been an adequate approach in the 1970s and early 1980s, it has long been understood that such an assessment of 'random hardware failures' alone represents only part of the picture. The growth of complexity over the last three decades has led to the dominance of 'systematic' failures, which cannot be predicted and assessed by quantitative techniques alone. IEC 61508 has established and codified the need for a raft of techniques and measures, throughout the lifecycle, to minimise these systematic failures.

Awareness of human factors

A particular benefit, which has arisen from the last 25 years' work in this area, is the understanding of the role of human factors in major incidents. A mass of empirical human error data has led to robust prediction models and a limitation on the degree of risk reduction claimed by manual responses to alarms.

Understanding of cost limitations (e.g. ALARP)

Although only covered as guidance in IEC61508, the practice of setting quantitative integrity targets has led to the concept of ALARP (as low as reasonably practicable). Because meeting a quantified target has become the object of an assessment, then the question arises as to by what margin.

The ALARP concept follows with the idea that further risk reduction should only be carried out until the cost becomes disproportional [2, 5], at which point, it is argued, additional resources are not justified and could more fruitfully be employed in risk reduction elsewhere.

Wider availability of quantification tools

The almost universal application of risk assessment has provided the market impetus for the development of a wide range of failure data and calculation tools. These greatly reduce the time and effort needed to carry out assessments and, therefore, the number carried out increases for the same amount of manpower.

Undesirable Outcomes

Inappropriate use of the 'SIL' term

Obsession with the 'SIL' word has grown amongst a very large number of people (including many so-called 'experts') without understanding its meaning. It is, in fact, only an arbitrary metric invented in order to classify the qualitative techniques and procedures throughout the life cycle, which are deemed to minimise systematic failures. Integrity targeting (not 'SIL targeting', as it is widely described) should establish maximum tolerable risks and failure rates as targets for the quantitative assessment.

'SIL' is a necessary and useful concept but only as a secondary consideration during the integrity targeting process.

Only because of the qualitative activities does it become necessary to have 'bands' of rigour instead of numerical targets. The choice of four SILs is again arbitrary and they might just as well have been labelled bronze, silver, gold and platinum. The impression of numeracy, given by the terms 'SIL 1' to 'SIL 4', is potentially misleading.

Nevertheless integrity studies are referred to as 'SIL' studies, integrity targets as 'SIL' targets, and so on. Sadly, this trend is worsening as the misunderstanding widens. This is not helped by consultancies and products seeking to incorporate the 'SIL' mnemonic into their titles.

Ascribing SILs to hardware rather than to functions

There has become an almost universal practice of describing every aspect of an instrumented loop and its procurement by means of the SIL. Although theoretically not wrong, it promotes the idea that a piece of hardware (and its software) has a safety integrity level. It does not. It is functions that have SILs, and the elements of a safety-related system need 'suitability' for use at a particular SIL and only in respect of a defined failure mode.

The plethora of misunderstanding embraces the idea that an item can have a SIL without any mention of how it might fail. It may fail in many ways, each of which relates to a different potential safety function. Since its rate of failing and proportion of hazardous failures will be different for each mode, it will potentially have a different SIL for each mode.

Dumbing down of targeting methodology

The spread of misunderstanding, emphasised in the earlier sections of this article, is largely due to a phenomenon which does not seem to apply to other disciplines within engineering.

That is, the obsession that everyone must have a 'say' and a 'part' in the safety assessment process. As a result, there has been a disturbing trend to dumb down the processes by creating pocket 'methodologies' that allow non-experts to replace experts. The most appalling example is the use of risk graphs, which enable amateurs to establish so-called 'SIL targets' with no need to establish failure mode details or proper quantified risk targets. Worse still, they, and other makeshift techniques, are so widely used and taught that it is possible to attend courses in their use and obtain certification, giving the impression of expertise in the subject without any proper understanding of the underlying principles and mathematics involved. Furthermore, as with other disciplines, experience gained over many years is vital in order to make effective judgments.

The author frequently encounters 'experts' who can neither explain the difference between a rate and a probability nor establish an appropriate maximum tolerable risk and calculate the maximum PFD (probability of (dangerous) failure on demand) required of a risk reduction function.

Certification and SIL application to too low a level

Misunderstanding of the SIL term, in its application to simple devices/components, has led to requests for it to be demonstrated at component levels. The only parameters, related to functional safety, for an electromechanical relay is its failure rate and the proportion of 'fail to open' and 'fail to close' modes. To ascribe a 'SIL' at this component level is both unnecessary and misleading. Any question of SIL relates to the safety function in which it is used. However, more complex items (e.g. field detectors) may claim a SIL capability based on safe failure fraction and design cycle rigour.

The author has been requested (on many occasions) to certify a SIL 2 capability for a sounder or beacon. Anyone with appropriate expertise knows that these components can only be part of a 'human response' function, which should never claim more than SIL 1: an example of lack of knowledge and of understanding.

Lack of focus in the choice of rigour

Having already covered the reasons why IEC 61508 [3] has been an excellent innovation in principle, it has nevertheless to be said that the document is extremely lengthy, verbose, repetitive and poorly structured. IEC 61511 [4], despite being couched in the usual lengthy style of standards, nevertheless achieves a great deal by way of a simpler approach.

This, together with the 'page by page' mentality of many users, often leads to slavish rigour to written clauses in the belief that this achieves a robust review.

It needs to be realised that (along with the ALARP principle) there is an optimum resource for any assessment. Thus a 'page by page' approach is in danger of losing overall perspective and of watering down the effort in areas where it matters. A robust approach involves an informed selection of key areas of criticality and then applying the assessment effort accordingly.

Proliferation of guidance

There seems to be a compulsion for bodies to write their own version of the standard. Most documents become a reiteration of the same text with slightly different terminology, headings and layout. Hence the nightmare of comparing vast quantities of guidance which essentially say the same thing whilst differing in respect of enormous quantities of trivial detail. This task of keeping up with the totality of 2nd tier guidance is therefore considerable but adds little to actual safety.

The Way Forward

Everyone to his own expertise

Industry (and its safety-related fraternity) should discourage the practice of non-expert participation and promote training with some academic content to qualifications. This must therefore include an understanding of probability and statistics and its underlying mathematics [6], and only persons with appropriate aptitudes, along with adequate experience, should seek to represent themselves as experts in this area.

Have one central standard

Industry should discourage the wasteful use of effort in writing guidance after guidance on a subject which is already over documented. This effort would be better employed improving the presentation of the existing standard and, also, in actually carrying out assessments.

Dr David J Smith (www.technis.org.uk) is also known as Technis. He is a consultant in risk and reliability as well as a well known author and provider of a suite of related software packages. Copyright (2011) in this article is held by Technis.

References
[1] Smith D J. Reliability, Maintainability and Risk, 8th Edition, Elsevier (Butterworth Heinemann) 2011 ISBN 9780080969022.
[2] Smith D J and Simpson KGL. The Safety Critical Systems Handbook (A straightforward guide to functional safety IEC61508) 3rd edition, 2010, Butterworth Heinemann ISBN 9780080967813
[3] IEC Standard 61508, 2010, Functional safety: safety related systems - 7 Parts
[4] IEC Standard 61511: Functional safety - safety instrumented systems for the process industry sector
[5] HSE. R2P2 Reducing Risks, Protecting People, HSE's decision making process, HSE Books, 2001
[6] www.technis.org.uk

Human factors — how little can you get away with, and how much is right?

Brian Sherwood-Jones

BAeSEMA

First published in Safety Systems 6-3, May 1997

Introduction

Ever since the World War II research into pilot error, it has been recognised that preventing operator error is important to the design and operation of safety-related and safety-critical systems. It is generally considered that 'human error' accounts for 60 - 80% of the risk exposure. Standards such as IEC 1508 are tightly bound to the `software system' rather than the 'work system' (see below), and as such will not address more than 20 - 40% of the risk exposure. This article addresses recent developments in Human Factors standards, legislation and research, to put together a synthesis of what might now be expected as 'best practice'. This best practice seems to be very wide-ranging, but it must be considered particularly relevant to safety-related and safety-critical systems. Safety-Critical Systems Club members are invited to comment on its scope and accuracy. The article is written from a technical point of view and cannot be construed as legal advice or formal interpretation of legislation.

Background

Reason [1] has proposed three 'ages' of safety concern: the technical era, the human error era, and the socio-technical era. In the technical era, safety measures were primarily directed at mitigating technical failures. Operator error was just that. In the human error era, numerical prediction techniques were applied to operator slips and lapses and it was recognised that man machine interface (MMI) design affected risk.

We are now in the sociotechnical era which recognises the importance of managerial and organizational issues and the need for safe systems of work. The need to provide assurance of a safe work system is now embodied in legislation and affects both design and operating organizations. This is a recent development arising from standards associated with the 'six pack' of EC Directives.

The last two years has seen a remarkable convergence in the definition of Human Factors best practice across a wide range of applications. In particular, the Provision and Use of Work Equipment Regulations are very wide-ranging. Ergonomic requirements are given in BS EN 614-1 (`Safety of machinery - Ergonomic design principles, Part 1: Terminology and general principles'. 1995), which specifies a set of design activities and criteria. The standard is much more wide-ranging and directive than its title implies.

Culture Change

The most problematic challenge faced by Club members is the culture change associated with moving from an engineering focus on equipment (with some consideration for the user when it comes to MMI design) to an integrated approach to a whole work system where 'good design starts with the operator'. A work system takes in the equipment and its context of use and embraces all the work equipment and its documentation, the workplace and the working environment. It is necessary to consider all users and their tasks, together with the management and organization.

Management

Integration between Human Factors and mainstream engineering is required, and the design process needs to allow for iteration and time for feedback. User involvement needs to be planned and implemented.

The key management task is to answer the question in the title of this article. Current best practice recommends a pragmatic, risk-driven approach. Unnecessary activity and documentation is discouraged, but the Human Factors work that is done has to be done to high standards.

Risk management needs to be done to an organized set of categories. Apart from consideration of physical hazards, the software community needs to take account of task design risks. The allocation of tasks to the users and to equipment needs to be documented on the basis of conscious design decisions, taking account of task priority and frequency. User skills, capabilities and experience need to be recognised in the allocation. Task design should avoid overload, underload, repetitiveness, time pressure and machine 'pacing'.

Setting usability goals (in terms of effectiveness, efficiency and satisfaction) is required, followed by the monitoring of their achievement.

The activities in a Human Factors programme can be categorised under the headings of analysis, design and test and evaluation. The minimum requirements for best practice in each of these headings is set out below.

Analysis

Risk and hazard identification and assessment are required. They should include analysis of the current situation, what actually happens, non-routine occurrences, all interactions, existing measures, and the effect of new technology.

User analysis is required. It should include analyses of all groups of people and should identify high-risk groups; it should establish specific user characteristics and identify different types of users.

Task analysis is required. It should include:

- Organizational goals;
- Allocation of functions for task design;
- User involvement. Workplace, workstation, and environment analyses should also be undertaken.

Design

A best practice design programme would require:

- Use of Human Factors
- literature and standards;
- Conformity with ergonomic principles and design intent;
- Iterative prototyping, the assessment of findings and feedback to the design authorities;
- User involvement;
- Establishment of training needs.

The Human Computer Interaction (HCI) should take account of the 'principles of software ergonomics' and other closely related requirements. The design intent here is that the dialogue is:

- Suitable for the task;
- Self descriptive, immediately comprehensible, clear and unambiguous;
- Error tolerant, with avoidance of inadvertent operation;
- Suitable for learning and easy to use;

- Adaptable, with provision for individualisation. The dialogue should provide:
- The minimum information and controls required, reflecting task priority and frequency;
- Controllable pace and direction;
- Consistent symbols and layout which conforms with users' expectations;
- Control and display compatibility;
- A match to illumination;
- Feedback on task performance.

Test and Evaluation

Best practice would expect a progression of prototyping, user trials, field trials and in-service evaluations against usability goals.

The challenge for the SCS community is to find cost-effective ways of providing the necessary assurance. Part of this is being able to measure the extent to which a design process has incorporated human-centred processes to address the work system as a whole. Suitable maturity indices are being developed in Europe and the USA and will be discussed in a subsequent newsletter.

As mentioned earlier, the most difficult task we face is integrating the Human Factors engineering processes and culture with the equipment engineering processes and culture so as to provide real systems engineering.

References

[1] Reason J: 'Management Risk and Risk Management: Research Issues', in 'Human Reliability in Nuclear Power', Oct 1989, IBC Technical Services.

Application of Formal Methods in a Commercial Environment

Guy Mason

General Dynamics UK

First published in Safety Systems 18-3, May 2009.

This article describes the way in which formal methods were applied to a commercial programme to help give an overall system safety level of SIL4 as defined by Version 2 of Defence Standard (Def Stan) 00-56. In order to meet this system goal, formal methods were selectively applied to the software element of the system. The form this selective application took is the focus of this article.

The product was only one part of a complex system and so, not only was it necessary to satisfy our immediate customer of the compliance of the system but also the end user's safety advisor.

Software development to a high integrity is a demanding process, which is strongly biased towards the use of formal methods. To provide a commercially viable solution was therefore challenging.

Prior experience in applying formal methods in software developments within the company had led to the following observations:

- The benefits of the use of formal methods in software development are widely reported on; however, the necessary skills reside mainly in academia and specialist consultancies. This gives rise to issues with applying it at the scale of a commercial application.
- The focus must be on the overall safety of the application rather than solely upon its correctness with respect to requirements. In previous applications this was achieved by partitioning across additional processing resources, which has the side-effect of increasing recurring costs. This can be problematic in a competitive commercial environment.
- There is a necessity to maintain a balance between domain experts, with a clear understanding of system issues, and mathematicians who understand

the formal development technology. Allowing either side to dominate is likely to cause problems.
- The largely unsubstantiated claim, that high integrity development reduces through-life costs, cannot be realised in an environment where the developer's follow-on business consists mainly of capability enhancement.

With these observations in mind, the programme was designed with the software residing on a single processing platform and with the safety-function software being limited in scope with a narrow focus. This restricted the scale of formal methods application, so as to minimise the impact of future changes on the overall safety of the system.

The architecture decided upon is shown in Figure 1. This is a basic control-and-monitor style of software architecture. Both the Executive and Safety Monitor components 'see' the same set of system inputs. The Executive software handles the main functional capability, with any action that results in the setting of a safety-critical output requiring a request to the high-integrity Safety Monitor. The Safety Monitor also performs autonomous control of safety-critical outputs, where necessary.

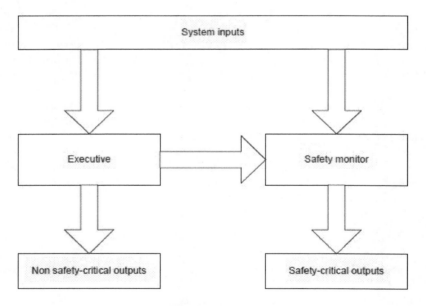

Fig. 1. Basic architecture

The Executive software was developed using a process that provided lower assurance of functional correctness than would normally be accepted for SIL4 software. The Safety Monitor software was developed using a process that gave high assurance of functional correctness, consistent with that for SIL4 software. The Safety Monitor and Executive software ran on the same processor. The focus of the software safety case, therefore, was on proving that the Safety Monitor formally met its requirements and that the Executive could not influence the correct functionality of the Safety Monitor implementation. Therefore, although the Executive did not require formal specification, there were still safety obligations on its software which required the performing of specific verification and validation tasks above and beyond those required for a non-safety-related development. The obligations were assured to a level of confidence consistent with SIL4 software.

The approach taken was:

- To identify the subset of the software that was safety-related (the Safety Monitor) and develop this to SIL 4;
- To develop the remaining software (the Executive) to a high level with respect to its possible effect on the correct functioning of the Safety Monitor - i.e. to provide an argument for its non-interference;
- To develop all software using a similar process but with some relaxation of independence and formal requirements in the Executive, the principal differences being the omission of proof activities and formal testing in support of the software safety case at system level.

System Level Safety Property Definition

The system-level safety properties were developed from the system hazard analysis and defined as invariants of the system, expressed in the Z formal specification language.

Consider the general abstract system depicted in Figure 2. It has inputs $I_1...I_m$ safety-critical outputs $O_1...O_n$ and state components $S_1...S_p$.

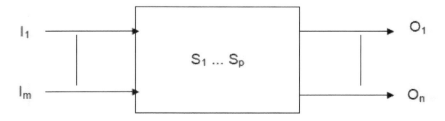

Fig.2. An abstract system

If a hazard is associated with output O_x being on (for example), then the safety property can be expressed as: $O_x = on => f(S,I)$ where f(S,I) is some logical expression involving the state and inputs. The formal safety property states that 'O_x may be on only if the condition f(S,I) on the state and inputs is satisfied.' Conversely, if the hazard involves the output being off, the safety property may be expressed as: $O_x = off => g(S,I)$

If there are hazards associated with both the activation and de-activation of the output, then condition g(S,I) must be the negation of f(S,I) (or otherwise the safety property does not account for all situations involving the two-valued output) and the safety properties for the two hazards may be written as the single safety property: $O_x = on <=> f(S,I)$ where the formal property states that 'O_x may be on if and only if the condition f(S,I) on the state and inputs is satisfied.'

The level of abstraction of the Z model was set high enough to be easily understood by domain experts with a little training in reading Z notation.

The partitioning of safety properties between hardware and software was made as part of the architectural design process. The upholding of any properties with an inherent liveness issue were applied solely to hardware, to simplify the design and validation requirements of the software development. The result was that, if at any point the software does nothing (e.g. due to a processor check stop), there is no safety concern.

Safety Monitor Software development

A functional specification for the Safety Monitor was also produced in Z, at a low level of abstraction, which allowed an almost one-to-one transition to the code. The sequential nature of processing and inherent latency in a real world system had to be taken into account here. The requirements imposed by the chosen system and software architecture were also apparent, with the design specifically detailing the necessary mechanisms for the software control of safety-critical outputs following requests from the Executive.

To achieve the kind of analysis requirements required for the software safety case, code was written in the SPARK subset of Ada, and the SPARK examiner toolset, developed by Praxis High-Integrity Systems was used for the analysis.

A UML tool was used to produce package skeletons with initial SPARK information flow annotations added at that point. The Ada bodies were then manually coded, with the SPARK examiner used to ensure that the code was in compliance with the subset and that the information flow annotations were consistent with the architectural concept of Figure 1. Annotations were then added to the code to facilitate partial correctness proof.

Executive Software Development

The Executive was designed using a standard UML development path, resulting in automatically produced package skeletons. As with the Safety Monitor, the procedure continued manually with the production of SPARK Ada code and information flow annotations. The difference here was that partial correctness proof annotations were not written.

Verification and Validation Activities

Partition Analysis
A two-step analysis was performed using the generated by the SPARK examiner tool, to ensure that the Executive could not systematically affect the safety-critical outputs:

- Safety-critical state identification - it was first necessary to identify all the inputs and state items upon which the safety-critical outputs were dependant and to ensure that they were implemented with SIL 4 functional correctness.
- Segregation Analysis - the annotations in the system were examined to ensure that safety-critical outputs were only generated by the Safety Monitor or through requests to it from the Executive software.

Further analysis was performed to confirm adequate mitigation against unintentional interference with Safety Monitor operation through overflows and processing resource denial.

Formal Proof of Safety Monitor Design
To ensure correct operation of the Safety Monitor software, both the Z specifications and the resulting code were subject to formal proof activities. These were as follows:

1. Formal proof that the functional Z specification correctly reflected the top-level safety property Z specification. The abstract invariants in the safety property Z specification were translated into equivalent representations defined in terms of the low-level specification's components and then checks were performed to prove that the functional specification upheld the translated invariants. However, the different scope of the two specifications, and the semantic step between them, resulted in a reliance on rigorous argument as opposed to formal proof in some areas.
2. Formal proof activities that the functional Z specification was self-consistent. This proof showed that definitions within the model were meaningful and that the model's state items could achieve values that satisfied all the defined constraints. It was essential that this task be completed successfully, as the conformance of an inconsistent model is worthless.

3. Safety Monitor partial correctness code proof. This ensued that the code correctly implemented the functional Z specification. This was done by using the SPARK examiner to produce verification conditions (VCs) from the partial correctness proof annotations. These VCs were subsequently proved partly by the automated simplification tool and partly using the interactive Proof Checker tool (both tools being supplied with the SPARK toolset).

Unit test activities were also performed on the Safety Monitor, using a test harness. These tested the upholding of the software-allocated safety properties by simulating aberrant behaviour from the Executive software. These tests were performed both in the development environment and on the target system.

Conclusions

A software safety case was formulated successfully, arguing an achievement of SIL 4 for the software design without arguing down compliance against the requirements of Def Stan 00-55. A critical part of that was achieved by utilising formal methods in a tightly focused area, showing that such methods can be successfully applied cost-effectively in a commercial environment. The use of a mixed approach to software development was successfully argued not to have compromised achievement of SIL4, based on the partition analysis work, to the satisfaction of the in-house independent approval authority, the customer, their independent safety authority, and the user's independent safety advisor.

The benefits of formally developed software were also evident with few instances of anomalous behaviour. Those that did occur were traced back almost entirely to inconsistencies in requirements. There was one tool fault, which was picked up by the additional unit testing activities, showing that diversity of verification techniques is still an essential approach.

Future application of the technique will benefit from addressing the following issues:

- The closing of the semantic gap between the high level invariant specification and the functional specification to reduce reliance on rigorous argument.
- The replacement of the tools used for the formal specification part of the development. Those used are now no longer supported and their replacement with new ones of the requisite strength is currently being investigated.

The company is continuing to use this selective approach to the inclusion of formal development technology in its highest integrity products, indicating that the commercial application of such techniques has real value.

A discussion of risk tolerance principles

Odd Nordland

SINTEF

First published in Safety Systems 8-3, May 1999.

1 Introduction

There is a number of approaches used to identify whether or not the risk posed by a given technical system is 'tolerable'. These are all attempts to define objective, rational criteria for determining whether or not enough has been done to eliminate risks in order to fulfil public expectations and demands. Where the risks cannot be completely eliminated, they should at least be reduced to such a low level that the general public would be willing to tolerate them and still accept deployment of the system. Let's first take a brief look at some of the currently used principles.

2 Currently used principles

The French GAMAB principle ('Globalement Au Moins Aussi Bon': globally at least as good) assumes that there is already an 'acceptable' solution and requires that any new solution shall in total be at least as good. The expression 'in total' is important here, because it gives room for trade-offs: an individual aspect of the safety system may indeed be worsened if it is over-compensated for by an improvement elsewhere. It is closely related to the British ALARP principle, as we shall see in a moment.

The ALARP principle (the residual risk shall be 'as low as reasonably practicable') is the only one described in IEC 61508, in Annex B to Part 5 of the standard. That annex is of informative rather than normative character, so it should not be concluded that the ALARP principle is the only one that is compliant with the standard.

The ALARP principle assumes that one 'knows' a level of risk that is acceptable to the general public and requires that the risk posed by any new system shall at least be below that level. How far below is where the term 'reasonably practicable' comes in: theoretically, an infinite amount of effort could reduce the risk to an infinitely low level, but an infinite amount of effort will be infinitely expensive to implement. So we have to identify a second level of risk that is so low that the public will accept that 'it's not worth the cost' to reduce it further.

Now, associating risk reduction with cost tends to be misunderstood. Of course, if achieving a safe system is prohibitively expensive, the system just won't be built. But cost is not just a matter of money.

Let's take a rather primitive example to illustrate this. If the risk of being involved in a train accident at high speed is unacceptably high, one way of reducing it would be to reduce the speed of the trains. This, however, would mean that the duration of a train trip would increase, so that the exposure time to the risk of a low speed accident would also increase. In other words, the cost of reducing the risk of high-speed accidents would be an increase in the risk of low-speed accidents. Now if that increased risk at low speeds results in an increase of the total risk, then the cost of reducing high-speed risks would be deemed too high.

Now this is effectively what the GAMAB principle says: if the increase of low-speed risk is less than the decrease in high-speed risk, the change is acceptable. So the GAMAB principle is basically the same as the ALARP principle!

The German MEM (minimum endogenous mortality) principle starts off with the fact that there are various age-dependent death rates in our society and that a portion of each death rate is caused by technological systems. The requirement is then that a new system shall not 'significantly' increase a technologically caused death rate for any age group. Ultimately, this means that the age group with the lowest technologically caused death rate, the group of 5 to 15 year olds, is the reference level.

In the CENELEC pre-standard prEN 50126, the reference mortality rate is given as 2.10^{-4} fatalities per person per year, and the limit for 'significantly' augmenting this rate is given as at most 10^{-5} fatalities per person per year, i.e. 5% of the reference value.

3 Differential Risk Aversion

In the Introduction it was pointed out that the various approaches are attempts to identify objective, rational ways of determining whether or not a risk is acceptable. Nevertheless, they have to take irrational, emotional factors into account.

People tend to be more willing to accept a risk if they think that they can directly influence how strongly it affects them. They are willing to accept a horrendous death toll on the roads, because they directly control the cars they drive. But for public transport, they are much more demanding: if they're going to put their lives into the hands of somebody else, then every precaution must be taken to protect them!

They also tend to view accidents singularly. If a single accident can cause a gigantic catastrophe, it will be much less acceptable than a hoard of small accidents that each have apparently minor effects, even if the total is much worse. When a thousand people get killed in a single accident, it is taken much more seriously than ten thousand deaths in fifty thousand road accidents spread over a year.

This effect is taken into consideration in most countries by introducing a 'differential risk aversion' (DRA). Basically, it is assumed that accidents up to a certain severity can be regarded as being equally serious, severity being interpreted in terms of the death toll. Above a certain threshold, people will react increasingly negatively, so their willingness to tolerate the associated risk will decrease accordingly.

In Britain and Germany, for example, a linear relationship is applied to the DRA, i.e. the decrease in risk acceptability is directly proportional to the increase in the potential death toll. This is not always expressed explicitly in all countries, whilst some are even more demanding. The Dutch, for example, use a DRA that is proportional to the square of the potential death toll.

From the above we can see that there is a risk level that is so high that people will categorically refuse to accept it. Such risks are intolerable. But there is also a level that is so low that people regard the risk as being negligible. The region between these two levels is where the tolerable risks lie, those non-negligible risks that people are willing to live with.

4 Determining tolerable risks

What sort of acceptance criteria should be used then? Let's start with a look at the railway sector, because it's an area where most people have a fairly clear impression of what kind of risk they're willing to tolerate!

Engineers and railway authorities tend to use accidents or casualties per passenger kilometre as their unit of reference for any comparison with other systems, be it a neighbouring country's railway system or other traffic systems. This quantity is certainly a very rational, calculable entity that does say something about traffic density and accident rates, but its use as a reference for acceptability is questionable. For manned space travel, the colossal distances involved result in exceptionally high figures for passenger kilometres and correspondingly low figures for the risk expressed as casualties per passenger kilometre. On that basis, space travel is an exceptionally safe affair, in strong contrast to the safety efforts of NASA, ESA and all the other space agencies!

From the point of view of the 'man in the street', it doesn't matter how far he got before being killed. He's more concerned about how often he can take a train without losing his life. So some combination of mortalities per accident and accidents per trip (not kilometre!) will interest him.

He also has a chronically short memory! If accidents occur once every ten years, even if they are full-scale catastrophes they will still be regarded in isolation. If accidents occur once a month, even if there are no mortalities at all, the system will be regarded as unacceptably unsafe.

There are differences between countries and societies. It was pointed out earlier that the Dutch, for example, use a stronger DRA than the Germans or British. And a look at the public attitude to traffic safety in underdeveloped countries shows that there is virtually no awareness of the risks involved in public or private transport!

Finally, it is also a political issue. When the population is highly 'tuned in' to a discussion of safety, the willingness to accept residual risks will decrease. On the other hand, if a technology is considered to be of vital importance, people will be more willing to accept the risks it entails. So the benefit provided by the system will influence peoples' willingness to accept the risk.

Thus we see that the tolerable risk level for a railway system must be some function of:

- the average number of casualties per accident, C/A;
- the average number of accidents per journey, A/J;
- the distribution of accidents over time, dA/dt;
- a differential risk aversion factor $f(C)$; and
- a factor b describing the benefit provided by the system.

The willingness to tolerate the risk will decrease when any of the above factors except b increases. It will increase with b, so assuming equal weighting of each of the factors, we can define tolerability T as:

$$T = b / (C/A * A/J * dA/dt * f(C)) = b / (C/J * dA/dt * f(C))$$

A high value of T means that people will in general be willing to tolerate the risk; it corresponds to a low risk level. A low value of T means something has to be done to reduce the risk; it corresponds to a high risk level.

Note that neither the total number of passengers nor the distance travelled (i.e. passenger kilometres) go into this equation! The average number of casualties per journey (C/ J) and the accident rate (dA/ dt) are measurable quantities. The differential risk aversion factor f(C) will not only be different for each country, it may even differ in different regions of a country. In addition, it will vary over time, reflecting changes in politics and peoples' thinking.

For simplicity, we will restrict ourselves to a single geopolitical region and set f(C) to 1 in the equation. The accident rate (dA/ dt) is of course dependent on what one calls an accident, which in turn is (indirectly) dependent on the definition of 'casualty'. Generally, the term casualty is restricted to mean any case where a person needs medical attention, i.e. we exclude environmental or financial damages but include non-mortal injuries, so an accident in our sense is any event where a railway system is the cause of a casualty. It is then possible to determine the accident rate as the number of such accidents per year and to identify the number of casualties from those accidents.

The number of journeys is not simply the number of different routes multiplied by the number of departures per route, because people may make several journeys over parts of the same route, for example. The number of journeys is equal to the number of tickets that are sold! (Free trips or forfeited tickets shouldn't influence the statistics noticeably.)

The above considerations were made for railway systems, but they can of course be applied to any other form of public or even private transport. For other technologies, we need an appropriate interpretation of the term 'journey'. In most cases, 'journeys' will correspond to operational time, but it should be pointed out that this includes downtime caused by accidents! Only planned pauses in operation (e.g. maintenance shutdowns) should be excluded. So 'casualties per journey' becomes 'casualties per potential operational hour', and the tolerability becomes:

$$T = b/ (dC/ dt * dA/ dt)$$

Note that the term 'casualties' can be extended to include environmental and financial damages without influencing the underlying relationship.

It was pointed out earlier that a high tolerability corresponds to a low risk level and vice versa. And since we usually talk about risk reduction rather than tolerability increase, it is more practical to refer to high or low risk levels rather than low and high tolerability levels.

So we simply invert the expression to get an expression for the risk level:

$$r = 1/T$$
$$= (dC/dt * dA/dt)/b$$

5 Conclusion

We have found an expression that relates risk levels with the casualty rate, the accident rate and the benefit provided by the system involved, regardless of what kind of system we're looking at. The higher the risk level, the less tolerable the risk will be.

There will always be a threshold risk level, above which a risk is considered to be completely intolerable, and risk reduction measures must always aim at getting the risk down below that threshold for unacceptable risks.

But there is also a lower threshold value, below which risks are considered negligible. Where these thresholds lie depends on social, political and geographical factors. Between the two thresholds we have a region where risks will generally be tolerated. Whether or not additional risk reduction will be demanded for such tolerable risks is ultimately a political question.

Quality assurance of software used in diagnostic medical imaging

Philip Cosgriff

Pilgrim Health NHS Trust

Boston, Lincolnshire

First published in Safety Systems 5-2, January 1996

Safety-critical systems will be found in many corners of your local acute hospital, but some areas are more safety-critical than others. Clearly, treatment devices are more safety-critical than those used for diagnosis because of the potential to cause direct harm to the patient. The software in some commonplace treatment devices (e.g., drug infusion pumps, cardiac defibrillators) is usually embedded and operates in real-time. This combination presents particular problems to the software developer, particularly in the area of testing. Since total reliability cannot be guaranteed, we have to rely on the designer to incorporate fail-safe features.

In diagnostic imaging (e.g., ultrasound, X-ray computer tomography, nuclear medicine) the situation is quite different. The computer processing of the raw data is usually performed off-line, and the output (processed images or derived quantitative parameters) will pass through at least three pairs of skilled hands before any decision is taken to treat the patient on the basis of the test results. That is, the scientist or technician who performed the processing, the specialist doctor (e.g., radiologist) who interpreted the images, and the clinician who referred the patient for the test. There is thus much more opportunity to spot errors but, of course, they can slip through.

The reason why we hear so little about such errors causing untoward effects is due in part to the general complexity of the patient management process. Treatment decisions are made on the basis of a variety of factors, only one of which will be the result of a particular diagnostic test. Even if inappropriate treatment is instigated as the result of a misleading test result, the patient's con-

© Philip Cosgriff 1996.
Published by the Safety-Critical Systems Club. All Rights Reserved

dition may not necessarily deteriorate to an extent that a problem is suspected. If it does, it still remains for the doctor to associate cause and effect - not easy with sick patients undergoing complex treatment (e.g., drug combinations, surgery) whose condition may have been worsening anyway. Only in cases where cause (inappropriate treatment) and effect are clear would an investigation result - which might then lead back to an incorrect test report.

A misleading test report can arise for two main reasons. The more likely is a simple reflection of the fact that diagnostic tests are not 100% accurate. In other words, we expect a proportion of results to be wrong (referred to as either false negative or false positive) although, of course, we don't know which ones at the time! Referring doctors realise the limitations of the tests they order and may therefore disregard a result in certain circumstances. The other main reason is a genuine fault of some kind: either human error (a mistake made by the scientist or technician performing the analysis or the radiologist doing the interpretation), or an equipment fault - which would of course include a software malfunction. This leads us to consider the steps being taken to ensure that our software does what it's supposed to, and does not contain serious bugs. The first thing to say is that the medical sector has lagged behind other sectors of the IT industry in the area of software quality assurance (SQA). Attempts are now being made to catch up by following the lead given by experts in the field. Current trends include adoption of generic quality management standards (ISO 9001 and ISO 9000-3), with sector-specific work being done in the areas of specification standardisation, end-product testing, and certification.

Adoption of the ISO 9001 standard is effectively being forced on commercial medical equipment manufacturers by the recent EC Medical Devices Directive. The Directive is basically a list of essential requirements (ER); the one relating to software (12.1) being a catch-all statement - lawyers at work? - which has been interpreted in some quarters as indicating compliance with ISO 9000-3 (or equivalent).

Belatedly, work is now being done on producing standardised medical device specifications - including the computer component - which should avoid software manufacturers having to implement different algorithms for doing essentially the same thing. As a result, the 80/20 rule - closer to 90/10 in my experience - is very much in evidence in most commercial medical software packages. However, user attitudes need to change for emerging standards to be widely accepted and adopted. The onus here is on the respective professional bodies (both national and European) to exert some influence over their members.

With regard to end-product testing, an interesting lead has been given by a European COST project looking at quality assurance of nuclear medicine software. The novelty is the use of real patient data for test purposes. In the past we have relied on 'hardware phantoms' (physical models) and mathematical phantoms (Monte Carlo methods), but both approaches have serious limitations when trying to simulate human pathophysiology! The project has also gone one step further by commissioning the collection of independently validated patient data. In other words, the patient imaged in the nuclear medicine department would have undergone other diagnostic procedures (e.g., X-rays, blood analysis, biopsy) to establish the presence or otherwise of a specific disease beyond reasonable doubt. It should be added that these patients were undergoing this battery of tests anyway, so were not additionally inconvenienced.

We call these clinically validated data sets 'software phantoms' - even though the term is something of a misnomer. So far, software phantoms have been collected in the areas of kidney, heart and brain imaging. The idea is that these test objects will be distributed to hospital departments (on floppy disks or CD ROM) with a request that they process the data in their normal way. The results will then be returned to the Coordinating Centre. If a department's software produced an abnormal result from a data set which was known to be from a normal patient (or vice versa), then there must be something wrong with the software. You're right - this is just external quality control. It isn't quality assurance, but you have to start somewhere!

Accreditation is now acknowledged as a key area of software QA and work is just starting in the field of medical imaging. Two related issues are currently being discussed: accreditation of hospital departments to undertake (in-house) software development, and accreditation of test laboratories to enable them to perform independent testing of the end-products. In both cases it is envisaged that only large teaching hospitals (i.e., those linked to universities) would have the necessary resources. Unless their activities were centrally funded (either at national or European level), these centres would presumably have to charge for their services to cover set-up and running costs. The test laboratories would perform a standard test procedure using standard test objects. In the case of nuclear medicine, the test objects would almost certainly be the software phantoms mentioned earlier. Conforming software would be awarded a certificate which, for legal reasons, would be carefully worded to indicate what had - and had not been tested. Several groups are collaborating to bring these ideas to fruition, including a sub-group of COST B2, CEN TC251 (Working Group 5) and the EAL (European Association for Accreditation of Laboratories). If our recent bid to set up a Project Team within Working Group 5 of TC251 is successful, we also hope to involve Brian Wichmann (NPL), who will be well known to many of you.

Concerns about legal liability (re: Consumer Protection Act, 1987) have made in-house developers more circumspect about embarking on software projects, with increasing emphasis now rightly being placed on helping commercial manufacturers to produce a better product in the first place. In my own specialty (nuclear medicine) much of the in-house development work to date has been concerned with 'plugging the gaps' in the supplied commercial software. Hopefully, there will be less need for this type of activity as formal communication (along the lines of so-called 'rapid application development') between users and manufacturers improves. The in-house work that remained would then be true 'leading edge', performed by groups of graduate scientists in large hospital departments with the skills and resources to develop software properly - and maintain it.

You will have noticed that very little has been said about aspects of safety-critical design (formal methods, etc.). This is for two reasons. First, we have to 'walk before we can run' and, second, much of the rigour of safety-critical design represents something of an overkill at the low risk (diagnostic) end of the medical software safety spectrum. At the opposite end, applications such as implanted regulators (e.g., cardiac pacemakers) and surgical robots are as safety-critical as you can imagine, so the forthcoming IEC standard (1508) should certainly be relevant here. Indeed, the Medical Devices Agency of the Department of Health may well issue guidelines to manufacturers when the standard is finally published.

Although medical software developers have been slow to embrace any notion of SQA, there is just a chance that process- and product-orientated QA activities might come together reasonably well over the next couple of years and, hopefully, we will then have made up some of the lost time.

COST (European Cooperation in the field of Scientific and Technical Research) is a research facilitating body within the Commission of the European Communities

There's No Substitute for Thinking

Tim Kelly

Department of Computer Science, University of York.

First published in Safety Systems 20-3, May 2011.

Recently, there has been much discussion of the current state of safety case practice, especially in the UK defence domain [1]. Commentators remark that safety cases have become too large and complex, and talk of shimmering prospects of 'more elegant', 'stripped down', and 'minimal' safety cases. Of potential concern, is the implication that there are elements of current safety cases that can be excised or simplified without impact on the core principles of safety-case practice. This article attempts to strip back safety-case practice to these core principles. At this core, lies critical thinking – reasoning and justification supporting and surrounding a claimed position of safety.

Back to Basics

Safety is defined as freedom from unacceptable risk of harm [2]. Reasoning about risk of harm must be the technical basis for any safety case. For a safety case argument to be risk-based requires that the overall claim of acceptable safety be decomposed into arguments that justify the acceptability of the risks posed by the identified system hazards. The argument must clearly describe the means by which each hazard has been adequately eliminated and/or mitigated, and cite the evidence that exists to support this claim. This is the irreducible core of a safety case. Whilst maintaining this rigid and singular focus, thought needs to be given to the way in which acceptable risk is defined and determined. Similarly, there is choice (and, therefore, thought is required) as to the possible forms of risk management argument (elimination and/or mitigation) that can be used and the forms of evidence that will be cited. The overall framework of a risk-based safety case is one that requires and encourages thought and choice amongst sets of possible strategies and items of evidence. A consequence both of this freedom, and our inability to establish purely deduc-

© Tim Kelly 2011.
Published by the Safety-Critical Systems Club. All Rights Reserved

tive arguments of safety (i.e. to prove safety), is that another level of thinking is also required: having stated clearly our risk-based safety argument (and how it is supported by evidence) we must critically assess the confidence we have in this stated position. To do this needs an inquiring mind, coupled with knowledge of the accepted limitations of forms of argument and evidence, and experience of the domain within which the safety case is being established.

Thinking throughout the Safety Case Lifecycle

Thinking about the reasoning and confidence of the safety case should not be limited to the end of the system development and safety lifecycle. Instead, thinking should take place throughout the lifecycle, for example in the following phases:

Inception – We need to think carefully about why a safety case is being called for (e.g. over and above existing forms of assessment). What is its aim? (Mere compliance with certification is a weak motivation.) What is the scope of consideration? Which stakeholders can be called upon for input and insight? Proceeding with a safety case without addressing such thoughts (and resulting decisions) can easily set a development on the wrong track.

During Development – We need to reflect upon the evolving safety case. Are the interim findings telling us that the case isn't viable in its current form? Should we modify the system design? Should we find alternate forms of evidence and argument (e.g. that will provide us with greater confidence)? We should stop and think about such issues before blindly pressing ahead with a 'doomed' case.

Pre-submission – Only the most naïve of safety-case developers would pretend that their safety cases are without flaw or omission. As discussed in [3], the safety-case authors should enumerate the 'known unknowns' (or *assurance deficits*) of the safety case, and justify the acceptability of these deficits. Not to do this can be criticised as only presenting one side of the story – to have approached the safety case only from a positive perspective ('Why is this safety case convincing?') without embracing the negative ('Why might this safety case be wrong or unconvincing?'). It could also be viewed as 'gaming' the safety case acceptance process – to 'wait and see' if a reviewer can spot these deficits and whether they raise any concerns.

In Review – To supplement any review by the safety-case author, *independent* review and experience is required. By working from a fresh viewpoint, independent reviewers of a safety case can spot things that the safety-case author will have failed to notice. In addition to confirming the reasonableness of the assurance deficits already identified by the author, they have the capability to identify further deficits, and bring a potentially wider experience-base to the challenge of identifying counter-evidence and argument. 'Unknown unknowns' may remain after this process, but the open-endedness of the exercise doesn't excuse us from the requirement for the search.

In Operation – Operation should provide further stimuli to thought about the safety case. A safety case submitted and accepted prior to a system being in operation cannot help but be based upon estimation and prediction of how well it will perform in practice. In operation we have the opportunity to review the predictions. Is the operational experience in line with them? Does it challenge our assumptions? Are operational failures revealing of flaws in our argument or evidence? Even if, ultimately, the operational evidence is found to be supportive of our case, we still need to have exercised these thought processes.

The Enemies of Thinking

There are practices and pressures that constrain and even prevent the forms of wide-ranging thinking that have been discussed in the previous section. Examples include:

Over-prescriptive standards – Standards writers or legislators can sometimes appear to have 'done your thinking for you' – i.e. they've already worked out the forms of argument and evidence required, the priorities of those arguments, and the means by which confidence can be judged. This can be seen by some to absolve them from further mental engagement in the process. The danger is that this 'pre-packaged' thinking may not be a good fit for a specific safety case. It may focus our attention and effort in the wrong place, and may leave important issues overlooked. This should be mitigated by an *intelligent* application of standards that attempts to engage with the rationale behind prescriptive requirements.

Group Think – This can occur in safety-case regimes where, through culture, training and practice, a group starts to think in a similar fashion – using shared and established idioms, and failing to challenge group 'norms'. It is a cliché, but we must be allowed to 'think outside the box' in safety-case development (possibly by utilising methods such as the one described in the next section).

Cost and Time Pressure – You may have been reading this article with the growing feeling, 'This is all very well, but I'm not given the time or money to think this way.' It is true that cost and time pressure, and commercial considerations, can often be the enemy of creative and critical thought processes. For example, the overwhelming commercial imperative to get a safety case accepted can lead developers to prioritise compliance arguments over more wide-ranging safety arguments, or can be viewed as a strong disincentive to be honest and thorough in the identification of assurance deficits ('What? Admit we haven't covered something? Are you mad?') There are no easy answers to the mitigation of this risk, other than to acknowledge the crippling influence such pressure can bring to bear on those who are *required* to think as part of their job, and to do our best to protect the time and money dedicated to these activities.

Safety Case Development with 'Hats On'

Some practitioners find the forms of critical and creative thinking discussed in the previous two sections 'second nature'. Others do not. Some may usefully consider deploying established thinking techniques, such as de Bono's *'Six Thinking Hats'* [4]. This approach is used to help individuals and groups look at decisions from a number of perspectives – identified by the white, red, black, yellow, green and blue hats. In the safety-case domain, the 'decisions' in question are our choices about the claims and strategies of the safety argument and selection of evidence. The following sections explain how each of the thinking hats can be usefully deployed in safety-case construction.

White Hat: Thinking with this hat calls for us to focus on the *data* and *facts* that are available. In the safety-case domain this can be interpreted as thinking of the evidence that can be brought to bear in the safety argument. For example, what analyses, tests, operational data, and operator experience can we identify that will help focus and root the safety argument? Safety-case arguments must not become abstract constructions, divorced from the reality of the available facts. At any stage in safety-case development, it is worth putting on the white hat to take an inventory of the currently available evidence.

Red Hat: When wearing the red hat, we are allowed to bring intuition, gut reaction and emotion to the thinking process. You may think that this dimension has no role in the 'rational' process of safety-case development. However, there are a number of ways in which it can be beneficial. Firstly, it is important to recognise where irrational biases (e.g. concerning a particular argument approach, type of evidence, or particular safety standard) may be influencing the process. Secondly, much as we wish safety-case documents to be purely technical, they are not always free from persuasive or emotive language. It is important to recognise that the manner in which arguments are described can influencetheir perception. Thirdly, intuition (especially from those with experience) can often be the first step – if explored – to revealing a rational objection or supporting point of view [5].

Black Hat: Black-hat thinking calls for us to look at the negatives. In safety-case development this can be easily interpreted as the call to look for the counter-case: to identify counter-evidence that undermines, rebuts or undercuts the positive position we are putting forward.

Yellow Hat: The yellow hat calls for us to think optimistically, to search for the positives. In the safety-case domain, this hat induces us to think such questions as 'What supportive evidence already exists or could be easily constructed?' or 'What are the positive safety features that already exist in this design that we could cite in the safety case?'.

Green Hat: The green hat represents creative thinking. In safety-case development, this is where we should look for a wide range of possible strategies (without allowing them to be quickly shot down), to look for new ways to provide arguments and evidence of safety that may step outside tradition or established approaches. Green-hat questions could be, 'Does the safety case *have* to look like this?' or 'Why couldn't we use a different safety-case strategy?'

Blue Hat: The blue hat is concerned with controlling the thinking process. Blue-hat thinking is usually applied by those chairing a safety-case brainstorming session. If the ideas on how to take the safety-case argument forward are running dry, the chair may need to encourage green-hat thinking. If the thinking so far has been overwhelmingly positive (yellow hat) then they may need to call upon the group to think about the potential challenges to their arguments and evidence by getting them to wear the black hat (e.g. to identify assurance deficits and counter-evidence).

Summary

Thinking, and documenting our thinking (for others to comprehend), forms an essential and irreducible core to the purpose and processes of safety-case management. Ultimately, the quality of a safety case can be directly related to the quality of critical thinking that has been applied throughout its lifecycle – at inception, in development, in review and in operation. However, critical and creative thought is not always encouraged or easy to apply in practice. There can be enemies of thinking – such as excessive time or cost pressure, and 'group think' – and practitioners can find it difficult to think from multiple perspectives. In such circumstances, tools such as the Six Thinking Hats can help structure the thought process. To return to the challenge identified at the beginning of this article, above all we must ensure that any attempt to 'scale-down' or 'optimise' the safety case process must not be at the expense of structured and systematic approaches to thinking about safety-case arguments and evidence selection.

References

[1] C. Haddon-Cave, *The Nimrod Review: an independent review into the broader issues surrounding the loss of the RAF Nimrod MR2 aircraft XV230 in Afghanistan in 2006*, The Stationery Office, ISBN 9780102962659

[2] ISO 11014: Safety data sheet for chemical products, 2009

[3] R. Hawkins, T. Kelly, J. Knight, P. Graydon, A New Approach to Creating Clear Safety Arguments, in *Advances in Systems Safety*, edited by C. Dale and T. Anderson, Springer, 2011

[4] E. de Bono, *Six Thinking Hats,* 2nd Edition, Penguin Books, 2000

[5] M. Gladwell, *Blink*, Penguin Books, 2006

Dr Kelly is a Senior Lecturer in the Department of Computer Science at the University of York. He has supervised a number of research projects on various aspects of safety, and is also Managing Director of Origin Consulting (York) Limited.

The new MISRA documents

David Ward

MIRA Ltd

First published in Safety Systems 17-3, May 2008

Introduction

Today's road vehicles depend on their electronic systems to control functionality and to deliver the expectations of customers, manufacturers and legislators for comfort, safety, brand differentiation and environmental efficiency. Functional safety is therefore important in the processes for the design and implementation of these systems.

MISRA has a long history of developing practical 'how to' guidance for those developing safety-related electronic systems in the automotive industry and beyond. Recently MISRA has released a number of new documents and this article provides an introduction to them.

A brief history of MISRA

MISRA began life in the early 1990s as a project under the UK government-funded programme in safety-critical systems (originally known as the SafeIT programme). This programme sought to address the challenges emerging in the various safety-related industry sectors, particularly since the work that lead to the creation of the international standard and 'basic safety publication' IEC 61508 was underway.

MISRA's first publication was Development Guidelines for Vehicle-Based Software, more popularly known as the MISRA Guidelines. This document provided guidance on the development of safety-related software in the context of automotive embedded electronic control systems. One principal feature of these guidelines was that they adopted many of the principles from the drafts of

© David Ward 2008.
Published by the Safety-Critical Systems Club. All Rights Reserved

IEC 61508, but with a particular emphasis on interpreting them for the automotive industry. The project decided against producing a clause-for-clause interpretation of the standard, since it was seen as less than beneficial to be closely linked to draft documents that as we now know would undergo substantial change before emerging in their final form as IEC 61508. Instead, the authors took a flexible approach (now recognized as a 'goal-based' approach) that would take into account emerging technologies in software development and 'future proof' the MISRA Guidelines. Although the MISRA Guidelines were published in 1994 with an expected 10-year life (order of magnitude, not a precise 'time out') they are still widely referred to as a reference document for automotive software development processes, even despite the potential emergence of ISO 26262 (see below).

The key 'goal-based' requirements of the MISRA Guidelines are expressed in its Table 3, which maps requirements for each (safety) integrity level (SIL) to the various processes of a generic software development lifecycle. In particular, under the process 'languages and compilers' there is a requirement for a 'subset of a standardized structured programming language'. In the 1990s there was a move from programming of automotive systems in assembler languages to high-level imperative languages – notably C. Hence there was an identified need for such a subset of C for automotive use and this led to the development of the first MISRA C subset, which was published in 1998. MISRA C has since been widely adopted for embedded software development in most sectors, including rail, aerospace, defence, and medical devices.

Subsequent advances in software development have led to the development of new MISRA publications. Several new documents were released in November 2007 and they are described in the remainder of this article.

MISRA Safety Analysis

Since the publication of the MISRA Guidelines, a number of issues related to the management of functional safety and to the process of performing hazard (or safety) analyses have emerged. Firstly, the MISRA Guidelines contained requirements to perform safety analyses; however, little detail on the actual process was given. These requirements covered:

- Preliminary safety analysis, which should be performed during the concept phase of a project to derive initial safety requirements (including an initial SIL);
- Detailed safety analysis, which should be performed as the development progresses to confirm that the safety requirements have been implemented correctly.

Secondly, there is often now a requirement to demonstrate more direct compliance with IEC 61508, particularly with the safety lifecycle.

Finally, work has been underway for some two years now to develop an automotive version of IEC 61508, to be known as ISO 26262. At the time of writing, this article ISO 26262 is the subject of a 'committee draft' ballot within ISO (the International Standardisation Organisation). Compliance with this new standard will inevitably require many companies to implement updated processes for functional safety management.

The new *Guidelines for Safety Analysis of Vehicle Based Programmable Systems* address these issues by providing:

- A framework for safety management compatible with the IEC 61508 safety lifecycle and the typical development lifecycles used for road vehicles and their components. This framework also provides a migration path to ISO 26262, particularly where updated functional safety management processes will be required.
- Guidance on processes for performing preliminary safety analysis and detailed safety analysis.
- A generic model of risk and of hazard analysis in the automotive context, with examples on how to define and calibrate the parameters used in hazard analysis, leading to the allocation of risk reduction requirements to programmable electronic systems.

MISRA Autocode

The state-of-the-art in automotive software development is to use model-based development, and to generate the production source code in a target language automatically from the models. These 'model-based development' paradigms effectively represent another form of programming language, although one that is based on graphical elements rather than textual statements in a traditional imperative programming language. Ill-defined and unpredictable features are among the reasons that led to requirements for style guides and subsets of imperative languages; these reasons may also be observed in regard to graphical programming languages. Although few standards and guidelines for safety-related systems (so far) refer explicitly to model-based development, the general requirements for language subsets can still be regarded as good practice.

Therefore in the context of safety-related systems, 'programming' guidelines and 'language' subsets are required in order to use these model based paradigms effectively. Future standards, e.g. ISO 26262 and DO-178C, are likely to include explicit requirements to be applied to model-based development.

MISRA therefore embarked on an 'Autocode' activity, which will ultimately produce a hierarchy of documents on the subject of model-base development and automatic code generation, including:

- General good practice for the use of graphical modelling languages;
- Guidance particular to specific graphical modelling languages;
- Guidance particular to specific production code generators;
- Guidance on meeting the requirements of the applicable language subsets (e.g. MISRA C) of the target imperative language into which the code is to be generated.

The first documents to result from this activity, published in November 2007, are:
- *Modelling style guidelines for the application of TargetLink® in the context of automatic code generation* (to be known as MISRA AC TL);
- *Guidelines for the application of MISRA-C:2004 in the context of automatic code generation* (to be known as MISRA AC AGC).

MISRA AC TL is a document providing guidance on modelling style and 'language subset' issues for a specific code generator, while MISRA AC AGC provides guidance on how to meet the requirements of each of the MISRA C rules in the context of automatically-generated code.

The general principle in developing these and subsequent documents in the MISRA AC series is to place guidance as high as possible in the hierarchy. Where a specific requirement has been identified, for example in MISRA AC TL, the aim is to abstract a higher-level requirement from it that may subsequently be applicable to other modelling languages.

Future work

MISRA has also been developing a subset of the C++ programming language, to be known (unsurprisingly) as MISRA C++. Indeed it may already have been released by the time this article is published.

It is evident that MISRA has several activities that may be broadly grouped together under the heading of 'languages'. Most of the work to date in developing language subsets, whether of C, C++, or graphical modelling languages, has taken a bottom-up approach, starting from known insecurities in the language and constructing rules to restrict or prohibit the use of such 'features'. During the development of MISRA C++, some rules were adapted from MISRA C, and some of the new rules developed for MISRA C++ will be adapted for the next version of MISRA C. Meanwhile the MISRA AC activities are endeavouring to abstract guidance to higher levels. Similar approaches to language subsets have been seen in Japan, where studies have taken a top down approach to deriving language guidelines, starting from quality attributes such as reliability, maintainability and portability. In all of these activities, some common themes are emerging.

MISRA is therefore currently considering whether it is possible to create a set of generic language guidelines that encompass some of these common themes and avoid duplication between the different subsets.

For further details

Further details about MISRA's activities and its publications are available from the MISRA website: http://www.misra.org.uk

David Ward is Technical Manager of Advanced Electrical Engineering at MIRA Ltd with particular responsibility for functional safety of electronic systems. He is the MISRA Project Manager and the UK Principal Expert to ISO/TC22/SC3/WG16 'Road vehicles – Functional safety'.

ISO DIS 26262
The new automotive functional safety standard

David D Ward

MIRA Ltd

First published in Safety Systems 19-1, September 2009

Introduction

In the early 1990s the author well remembers attending a Safety-Critical Systems Club meeting where, during discussions, a delegate criticised the way software was developed in the automotive industry, implying it was amateurish. As recently as June 2009 there were similar comments on the Safety-Critical Mailing List, where someone said, 'The auto industry avoids like the plague having any of its electronic systems classified as 'safety-related systems' because then they must get SILs (safety integrity levels) and no one knows how to do that except in a fashion which has little apparently to do with safety. So all their systems, which aid steering and braking and stability, are 'driver assistance systems' so as not to be (safety-related systems)'.

In fact the automotive industry has a good record in terms of adopting best practice for the development of safety-related electronic systems. As early as 1994 the MISRA (Motor Industry Software Reliability Association) Guidelines were published, which were the first time the principles of the emerging standard IEC 61508, including the use of SILs, were interpreted for the automotive industry - some four years before the standard itself was published! Then, in 2007, the MISRA Safety Analysis Guidelines were published, based on more than 10 years' experience gained working with the 1994 Guidelines and with the direct application of IEC 61508.

© David D Ward 2009.
Published by the Safety-Critical Systems Club. All Rights Reserved

More recently, work has been going on to develop an international standard interpreting IEC 61508 for the automotive industry. The new standard, ISO 26262 Road Vehicles - Functional Safety is now publicly available for comment as a DIS (draft international standard).

This article gives a brief introduction to ISO DIS 26262, how it addresses the issues with applying IEC 61508 directly, and some of the challenges that users of the standard may well face.

Motivation for ISO 26262

Although IEC 61508 is designated as a 'basic safety publication' and a generic standard, intended to be used both as the basis of developing sector-specific standards and for direct application where no such standard yet exists, there are a number of well-documented difficulties in applying it directly to automotive safety-related systems.

These include:

- The standard was developed for low-volume not mass-market products;
- The safety lifecycle assumes installation then overall safety validation, whereas in the automotive industry prototype products are validated and then enter mass production;
- The methods for risk analysis are informative and industry interpretation is always needed;
- Subcontracting of work, a common feature of the automotive procurement chain, is not addressed;
- The safety functions are often considered to be distinct from the control functions, whereas in automotive systems the control and safety functions are usually inseparable;
- Human factors issues are only briefly addressed, whereas in vehicles the driver is an important part of the control loop;
- The 'techniques and measures' specified are quite prescriptive, orientated towards the process control sector, and in many cases are supported by references to documents that are quite historical.

Structure of ISO 26262

The DIS version of ISO 26262 consists of the following Parts:

    Part 1: Vocabulary
    Part 2: Management of functional safety
    Part 3: Concept phase

Part 4: Product development: system level
Part 5: Product development: hardware level
Part 6: Product development: software level
Part 7: Production and operation
Part 8: Supporting processes
Part 9: ASIL-oriented and safety-oriented analyses [ASIL stands for 'automotive safety integrity level]
Part 10: Guideline (informative)

There are over 350 pages and more than 550 requirements in the standard.

Objectives of ISO 26262

The developers of ISO 26262 have stated the following objectives for the standard:

- Maintain 'comparability' with IEC 61508 for product liability reasons;
- Adapt the safety lifecycle to the typical automotive development and operation phases;
- Refer to the milestones and prototypes or samples of typical automotive development processes;
- Include requirements for relationships between manufacturers and suppliers, and for distributed development processes;
- Adapt the hazard analysis and risk assessment for typical automotive use cases;
- Allow for application of typical automotive validation tests including hardware-in-the-loop, whole vehicle simulation environments, fleet tests and user-oriented tests.

In these respects it can be seen that ISO 26262 is indeed appropriate as an adaptation of IEC 61508 for the automotive industry, as these objectives address many of the issues raised above.

A key concept in ISO 26262

A key concept in ISO 26262 is the means by which safety requirements are derived and allocated to an electronic system or parts of the system, called 'elements' in the standard.

As with IEC 61508, a process of hazard analysis and risk assessment is called for early in the lifecycle of the item being developed. The ASIL (automotive safety integrity level) results from this analysis and, as with IEC 61508, indicates the risk reduction required from the associated electronic item. In contrast to IEC 61508, the ASIL does not include a probabilistic requirement for random safety integrity, although there are requirements relating to the violation of a safety goal due to random failures of the hardware.

The ASIL is assigned to a safety goal, which is a top-level statement of a safety requirement that results from the hazard analysis and risk assessment. For example, in an airbag system, one hazard is 'unwanted deployment' and the associated safety goal, to which the ASIL is assigned, could be 'prevent unwanted deployment'. Although in this example a safety goal that is NOT(hazard) has been stated, in many cases a key part of the safety goal, and of the subsequent safety concepts (see below) relates to the ability of the system to detect faults and warn the driver accordingly. Note that an item may have several safety goals.

Initial design of the item leads to specification of a functional safety concept, which is a statement of functional safety requirements to meet the safety goals; in particular how to manage the safety of the system in the presence of one or more faults. This is then further refined into the technical safety concept, which specifies how these functional safety requirements will be achieved in the design.

As the design progresses, the ASIL inherited by elements that constitute the item may be reduced by a process called 'ASIL decomposition'. This decomposition process, which applies only to the systematic aspects of safety integrity, may only be applied if the elements are architecturally redundant; that is, a failure in one of the elements will not in isolation lead to a violation of a safety goal.

'Techniques and measures' in ISO 26262

A key improvement in ISO 26262, compared to the application of IEC 61508 to automotive environments, is that the 'techniques and measures', generally known as 'methods' in the standard, have been made more goal-orientated, particularly in the software part of the standard. Thus, although the standard still contains detailed tables, which in many ways reflect those of IEC 61508, the emphasis is on demonstrating that the selected methods (whether those listed in the standard, or alternatives) comply with the corresponding requirements. Thus an important feature of ISO 26262 is that the emphasis for 'techniques and measures' is on making and documenting informed decisions, not merely 'box ticking'.

Potential issues in applying ISO 26262

What are some of the issues that may be experienced in applying ISO 26262?

Firstly, as will be well known to regular readers, IEC 61508 is a 'basic safety publication' and, despite its origins in the process control industry, it is intended that other industries should use IEC 61508 as a starting point and interpret it for their sectors. One key implication of this is that if an element is developed to a specific SIL in one sector, then it should be possible to carry this across into an application of the same SIL in another sector that also uses a version of IEC 61508. Although the risk may be different, the level of risk reduction (expressed as a SIL) is meant to be the same. However, in ISO 26262 the SILs (known as ASILs) are different and there is not a one-to-one mapping between the four ASILs of ISO 26262 and the equivalent SILs 1 to 3 of IEC 61508. Furthermore some of the detailed requirements are different.

Secondly, it was decided in the first edition, at least, to exclude trucks and buses from the scope; therefore the standard is initially only applicable to passenger cars. However, some of the systems and 'elements' are often common to passenger cars and sometimes to other applications away from road transport, so if these have been developed with an ASIL for passenger car use, how will this be related to a SIL for other applications?

Thirdly, the hazard analysis and risk assessment process is quite focused on European products and European markets, and in fact contains an implicit calibration on this basis. SIL 4 was explicitly excluded 'for product liability reasons' even though some companies have already applied IEC 61508 directly. The scope of application of the hazard analysis and risk assessment process assumes independent vehicles (i.e. not considering co-operative applications) and that the driver is always in final control. There are also no provisions for links to infrastructure systems where, again, IEC 61508 has already been applied directly.

All of these points suggest that attempting to apply the standard in a cross-sectoral context, or to some of the novel applications that are already being actively researched, such as co-operative systems and even autonomous vehicles, may be difficult.

Conclusions

ISO 26262 is clearly a significant development in automotive functional safety. In general it is to be welcomed as providing a sector-specific implementation of IEC 61508 that addresses many of the well-documented issues that will be encountered in applying that standard directly in the automotive context. Nevertheless, the standard (and particularly its approach to hazard analysis and risk assessment) is firmly based on the current practices and today's 'state of the art' of vehicle electronic systems, and makes few if any concessions to the novel applications that are almost upon us.

Review: IRSE Guidance on the Application of Safety Assurance Processes in the Signalling Industry (May 2010)

Andrew Rae and Mark Nicholson

University of York

First published in Safety Systems 20-1, September 2010.

General

The subject publication of the Institution of Railway Signal Engineers (IRSE), referred to hereafter as the 'IRSE Guidelines', is a response to difficulties which have been generated by the international nature of rail business and the move to 'open' systems rather than 'closed' systems operated by a single 'railway authority'. In particular, there is a conflict of philosophy between European Union legislation for interoperability certification and United Kingdom case law, legislation and regulations. A perceived consequence of this conflict is safety activity that involves unnecessary cost and stifles innovation.

The practical considerations raised by the publication ring true from both the perspective of industrial experience and an academic view of system safety engineering. Standards and regulations originally designed to help suppliers deliver safer systems can, over time, become onerous and unhelpful. This is particularly a concern when disproportionate power is given to a single individual such as an Independent Safety Assessor or a representative of the regulator, who may interpret the standards in a manner sub-optimal for both safety and competitive business.

Even good up-to-date standards can be misapplied. Project Managers and Safety Engineers alike should be concerned by inappropriate application of safety regulations. Poor system safety practice can harm the culture of an organisation by creating a perception that safety is a quality-assurance function and a hindrance to 'real' engineering.

As a directed response to perceived problems, the IRSE guideline is a good discussion document, but should be read as an adjunct to existing guidance and

© Andrew Rae and Mark Nicholson, 2010
Published by the Safety-Critical Systems Club. All Rights Reserved

regulations, not as stand-alone guidance. The proposed solutions are well motivated, but do not universally reflect good system safety engineering practice.

The Principles

The document defines fourteen guiding principles. For each principle we set out a summary, as described in the guidelines, along with an assessment. This assessment is necessarily subjective, but is informed by the problem the principle addresses and our understanding of current good practice.

Principle 1

Summary
This principle deals with early planning of safety. Planning includes both determining activities to be conducted and involvement of external parties and reporting lines between organisations. The intent of the principle is to avoid unplanned work, which is 'the major driver of cost overruns'.

Assessment
This principle deals with the risk of unplanned changes to safety activities. These may arise internally to a project if sufficient safety evidence cannot be generated. Of greater concern is conflict between the project's plans and an external party's expectations. If the external expectations are not communicated or understood until late in the project, significant unplanned work can result.

There is little to disagree with in this principle as stated. Its proper implementation requires recognition that the set of evidence needed to demonstrate safety can only be determined in the context of an outline safety case. This case will include both the main hazards and the design and assurance strategies for addressing them. Thus, the principle should explicitly cover early hazard identification and an outline of the safety case as part of the planning process.

Principle 2

Summary
This principle explicitly rejects the idea of an Independent Safety Assessor (ISA) as the 'default mode' for a project. The principle is based on assertions that competency of the design team is both necessary and sufficient for ensuring safety, and that ISAs have insufficient specific knowledge of the design to make reasonable judgements.

Assessment

This principle explicitly views the existence of an external ISA as a project risk. The reviewers believe that this principle may have arisen from experience with ISAs who are perceived to have acted beyond their legitimate role. Attempting to resolve issues with unreasonable ISAs is not an uncommon experience. For the external review to be realistic and helpful, an ISA familiar with railway equipment and operations is required, but the ISA should be reviewing the evidence of safe design, not the design itself. It is unfortunate that regulatory regimes, based on either certification authorities or independent assessment, may have a single point of failure located in the individual making an external judgement of a safety process. Whilst the role of an ISA is not intended to impose design or evidence decisions, in practice there is no available recourse should an ISA exceed their scope.

There is an important role for independent assessment as a checking mechanism for the design and execution of the safety process itself. This can be useful for commercial risk management as well as for safety. External review and endorsement of the safety process mitigates the risk of last-minute objections, and can help ensure that documentation is ready for regulatory review.

The reviewers believe that the risk of unhelpful ISAs can be addressed without dismissing the importance and helpfulness of a good ISA. A possible alternate principle would be to call for an adjudication process where an ISA and a project cannot find a mutually agreeable path forward. As with other parts of a safety program, the degree of ISA independence can be tailored according to the size, criticality and novelty of a project. The requisite independence can often be achieved by an internal ISA not associated directly with the project. As project engineers gain competence in performing safety activities, specialist safety engineers within a company can move towards an ISA role, and the IRSE Guidelines would do better to support this model than to reject the ISA concept.

Principle 3

Summary

This principle can be interpreted in a self-contradictory way, and so the appropriate text is quoted directly here:

'Whilst formal safety techniques such as hazard identification and closure are very important, these are not the only methodologies available as part of good engineering practice; engineers should carefully consider and use the most appropriate tools and techniques applicable to the design they are producing. Design review, including review by competent and experienced peers, remains a powerful process and is commended as good practice, particularly for complex designs. It is very important that safety analysis is performed, and the necessary supporting evidence is produced, contemporaneously (i.e. at the same time as the design work) by the same knowledgeable team, not generated later by a separate team.'

Assessment

This principle raises important considerations, but can be interpreted as suggesting that safety techniques such as hazard identification can be discarded. If the principle is read as suggesting that safety activities are ideally performed within the project team rather than by separate 'safety experts', this is a good principle. Experience with the demands imposed by signalling project timescales suggests that a review process independent of these pressures may be necessary to facilitate this principle.

A shortage of engineers with competence in both signalling systems design and selecting and performing safety activities will be a constraint in the short term, but the principle implies a growth in safety competency of all engineers, which would be a good thing.

Principle 4

Summary

This principle indicates that responsibilities should be clear, with no overlapping responsibilities or gaps. It also indicates that a single chain of responsibility is desirable.

Assessment

This principle directly conflicts with the existence of independent reporting channels for safety issues. A single chain of responsibility allows any decision-maker in that chain to prevent escalation of a safety concern. Since individuals in the chain will have competing responsibilities for safety and other project performance measures, this is a dangerous recommendation.

Principle 5

Summary

This principle states 'Project safety requirements and acceptance criteria should be agreed early in the project lifecycle.'

Assessment

This is a well-stated and important principle. It is very important for suppliers to know in advance not just what is required, but whether the client considers a particular type of evidence suitable for meeting the requirement.

Principle 6

Summary
This principle highlights the need for a well-defined change-control process, which includes consideration of impact of changes on safety management.

Assessment
There is nothing controversial in this principle. Change management is important for more than just safety, but certainly it has an impact on the success and cost-effectiveness of safety management.

Principle 7

Summary
This principle deals with architectural design, and subsumes a number of different issues. These issues include:

a. The need for clearly defined interfaces and boundaries;
b. System partitioning into understandable sub-systems;
c. Preference for open interfaces rather than project-specific solutions;
d. The role of existing infrastructure or retained element performance in achieving safety targets for upgrade projects;
e. Selecting architectures which group safety functions into a small number of sub-systems, which are decoupled as much as possible from the rest of the system
f. Avoiding setting safety targets for non-safety-related components 'just to be on the safe side';
g. Encouraging cross-acceptance of approved components.

Assessment
Most of these issues are non-controversial and require no further discussion here. Issue 'd' deserves highlighting as an important risk factor for upgrade projects. If the performance of an upgraded system depends on the performance of existing components, early agreement on how existing performance is to be measured and included in the assessment is vital.

Issue 'e' is a response to projects where all the elements of a system need to be developed to the highest safety integrity level (SIL). The idea of dedicated safety functions in an isolated module is frequently unhelpful, particularly in the context of a domain where availability and operability make contributors to safety. This principle should be carefully aligned with moves away from a single fail-safe state, as outlined in Principle 11.

Cross-acceptance is desirable in principle, but the differences in regulatory practice are beyond the 'minor differences of approach' that the article alludes to. Unless acceptance covers not just conditions of use, but also the specific safety functions or requirements which are met by a component or sub-system, it will be difficult to show that the prior approval is applicable.

Principle 8

Summary
Principle 8 expands the discussion of cross-acceptance, encouraging the use of the EU framework for cross-acceptance, and the use of notified bodies for conformity assessment.

Assessment
Whilst cross acceptance may be appropriate for individual items of equipment, with the prior acceptance used as evidence that an equipment item meets certain requirements, the issue of cross-acceptance is too complex to be stated as a principle for universal application.

Cross-acceptance is an important goal, but the first step should be to remove the concept of acceptance as a binary state (approved / not approved) and to attach clear scope and application requirements to each approval.

Principle 9

Summary
This principle draws equivalence between Safety Assurance and Quality Assurance, and asserts that independent checks can reduce quality.

Assessment
System Safety Engineering is not 'just a specialised branch of quality assurance' and any such implication is damaging to efforts to encourage highly skilled engineers to enhance their competency within the safety engineering discipline. Competence in design activities is not a substitute for systems engineering processes that provide appropriate requirements and ensure the existence of appropriate evidence that the finished artefacts meet these requirements. In the end it is the safety of the product that a user is interested in. This is not just a matter of process activities.

Principle 10

Summary

This principle notes that early involvement of stakeholders is important, but states that 'those with a marginal interest should not wield disproportional power, nor be allowed to add disproportionate costs at the expense of benefits to others'.

Assessment

This principle is the only one stated in a negative rather than a positive manner. It is clearly directed against a particular existing situation, policy or behaviour, but since it does not identify the target specifically it would be inappropriate to comment on the validity of the principle. Of course no party should wield 'disproportional power', but deciding how much power each stakeholder should have is a subjective judgement.

Principle 11

Summary

The eleventh principle states that safety must be considered in conjunction with operability. It observes that where safety is placed ahead of operability, this may ultimately lead to reliance on less-safe operational controls during periods of unavailability.

Assessment

There is little contention that technical safety and operability need to be considered together. This is good practice.

There is historical evidence supporting the conclusion that operation in 'degraded mode' is a significant risk factor for railways. Hence, availability of the full signalling system functionality directly contributes to safety. 'Fail-safe' designs that decrease availability may rightly be considered to be unsafe, if the consequence is extended periods of manual working.

This is a good argument for a safety case which makes explicit decisions about allocation of risk between hazards with competing mitigations. Such decisions would lead to availability and reliability requirements with specific targets, avoiding late competition between safety and availability.

Principle 12

Summary

This principle deals with the apparent paradox of safety systems which have had delayed introduction due to safety concerns. The principle asks for a 'whole of life' consideration, and implies that the potential benefits of a system should at times take priority over meeting specific safety targets.

Assessment

It is certainly possible that excessive regulation can impede safety innovation. It is also possible that new systems, if not appropriately managed, could introduce unacceptable risk. The principle is appropriate so long as both of these concerns are addressed in a balanced fashion. Resolving such issues is the role of government acting through regulatory bodies. Whilst it is appropriate for an industry body to point out that they feel current regulation is not achieving this balance, the IRSE Guidelines should make it clear that this is not a judgement that projects should be making for themselves.

Principle 13

Summary

Principle 13 covers the appointment of Independent Safety Assessors. It points out that a single ISA with the right set of competencies is preferable to multiple ISA organisations. It implies that contractors should not be required to appoint ISAs.

Assessment

The reviewers interpret this principle as an extension of Principle 2. The Guidelines are not in favour of current ISA processes, and view multiple ISAs with conflicting judgements as part of the problem. A single principle which sets out IRSE's view on when and how the use of an ISA is appropriate would be more constructive than the piecemeal criticism made in the Guidelines.

The idea of single ISAs is good, but not necessarily achievable when cross-acceptance of components and subsystems is involved. Systems approved at different times, or as part of other projects, would most likely be assessed by different ISAs.

The appointment of ISAs, and making decisions on their reporting responsibilities, are difficult issues. Whoever appoints the ISA benefits from retaining control of ISA competency, confidentiality, and the definition of the ISA's work. A contractor who appoints an ISA can use the ISA's report for multiple customers. Whilst the principle is not inappropriate advice for a regulator or client, it is not necessarily in the interests of a system developer.

Principle 14

Summary
The final principle indicates that safety processes should be tailored and scaled proportionate to the magnitude of the project and the risk.

Assessment
This is an important principle. 'Risk' is not known in advance, so perhaps this principle should be expanded to include elements of safety-criticality and project novelty.

Conclusion

The IRSE Guidelines make an important contribution to the debate about appropriate safety regulation and practice. They take a novel approach in that they focus on the practical effects of regulation and the important contributions that non-safety specialists make to system safety. Though addressed to a broad audience, they should be viewed as an open letter to policy makers rather than as a set of guidelines suitable for application by railway operators and equipment suppliers.

The specific principles are at times problematic, and appear as a response to current problems rather than a reasoned policy position. This is particularly clear in Principles 10 to 13, each of which focuses on the importance of one aspect of a trade-off decision. A more balanced discussion of the need for a trade-off, and the reasons for believing that the current policy position fails to make the trade-off appropriately, would result in a position that these reviewers would be happier to support.

The underlying philosophy of the IRSE Guidelines advocates a shift from systematic management of safety to individual responsibility and competence. This is contrary to an existing trend to manage safety by addressing organisational factors which lead to accidents. Such a philosophy is not necessarily wrong, but should be recognised as controversial.

Dr Mark Nicholson and Dr Andrew Rae are Research and Teaching Fellows at the University of York. Mark is the Program Co-ordinator on the Masters of Safety Critical Systems Engineering and has fifteen years experience of involvement with safety critical industry projects. Andrew was project safety engineer for numerous railway projects in the Asia-Pacific region, and has acted as Independent Safety Assessor for railway projects.

Complexologistification

Rob Collins

Sentient Systems Ltd

This article was first published in Safety Systems 7-1, September 1997.

A reply (immediately following this article) was published in 7-2, January 1998.

What is complexity?

I often attend learned conferences. Being of a somewhat mischievous disposition, I have occasionally been known to corner the presenters at such conferences and ask them 'dumb' questions. This essay is about one of the dumb questions that, more than most, I like to ask. Since I rarely listen to lectures these days that do not include phrases such as 'complex, safety-critical system', or just 'complex system', I have occasionally asked 'What does "complex" mean?'.

To be fair, most people I have asked have made a fair attempt at answering this question. The answers have concentrated on the type of associations that one might find in a thesaurus: complicated; convoluted, intricate, perplexing, elaborate, involved, etc. At least the connotative meaning of the word 'complex' seems fairly certain.

Not, however, being one to be so easily dissuaded, I have been known to continue 'But surely aren't all of these words just tautologies? What does "complex" actually mean?'. Definitions are refined to include, for example: 'consisting of many parts'; 'consisting of many connected parts' and so on. And I ask 'But if I made up a huge, regular grid of dots on a piece of paper, and even if I connected the dots with lines, nobody would regard that as really complex'.

Definitions then start to become more involved: 'consisting of many interacting parts', so that the notion of behaviour starts to play a role in the definition.

© Rob Collins 1997.
Published by the Safety-Critical Systems Club. All Rights Reserved

And even if I let it pass that there seem to be many things which are 'complex' and yet do not have 'behaviour' (in the sense that a machine or a biological system has behaviour), I could imagine a thing with many interacting parts that would also be simple. My favourite example is a Newton's Cradle of many thousands of balls that oscillates with perfect predictability and monotonous regularity.

Even authors whose text books depend on an understanding of the term 'complex' seem to find it hard to fathom a good definition. For example, readers of 'Dealing with Complexity', a text on systems science by Flood and Carson (1988), are left to infer what 'complexity' actually means from a series of observations such as 'complex situations are often partly or wholly unobservable'.

Some Definitions

Without more ado, let me share some of the less tautological answers from the literature that I have discovered. One definition that has been cited frequently is that due (independently) to Kolmogorov (1965) and Chaitin (1966). This definition relates to the 'complexity' of binary strings of digits, but may be generalised to other cases. The definition states that a measure of complexity is the length of the shortest algorithm that can generate a given pattern. If the pattern is not complex (for example because it is short, or has a repeating structure) then it can be proceduralised into a short program. If, on the other hand it is complex, then the shortest algorithm will, by definition, be no shorter than the string itself. This definition has a certain, unarguable, mathematical beauty. But it is also somewhat unsatisfying. The shortest algorithm to compute a given string cannot, in general, be determined, and thus, as a measure of complexity the definition is not helpful. Those interested in this general argument should refer to Casti (1994) and Chaitin (1966).

Bennett (1990) has gathered a diverse collection of definitions of the term 'complexity' applicable to physical and biological systems. These definitions refer to such properties as high free energy, the ability of a system to be programmed to perform like a Universal Turing Machine, the existence of long-range order in the system, and 'Thermodynamic Depth' — the amount of entropy produced in the system's evolution. Each of these is theoretically sound but again of little practical value to the engineer who needs a usable metric for complexity.

Of course, in the field of Software Engineering there is a proliferation of measures of computer program complexity. Few readers of this newsletter will be unfamiliar with measures such as those due to McCabe (1976) and Halstead (1977). But what do code complexity metrics actually measure? When they count the number of lines of code, or decision points, or iterations, or (in newer metrics) the depth of inheritance of a class, their definition is clear and precise.

But presumably the commodity that we are actually interested in is the relationship between such definite numbers and the difficulty of design, debugging and maintaining program code. Such a relationship with the cognitive complexity of such tasks is a wholly more elusive thing.

There is only a tenuous relationship between that which humans find easy or difficult and the measurably simple or complex. The game of table-tennis is 'simple' to play and yet requires mammoth computational resources; resources that have only recently been possessed by machines. Extracting square-roots of 35 digit numbers is computationally trivial and yet difficult for most humans without aid. As I type, Iain listening to Tanis' Loquebantur variis linguis, and I notice that sleeve notes describe the music as 'complex' (Gimell CDGIM 006). To my ear the music sounds rather 'pure' and 'uncomplicated'. The information content of the CD is presumably comparable with other music CDs. But what of the algorithmic complexity required to compose such music? That such problems are even conceivably approachable by algorithmic means is still open to debate.

In the field of safety-critical systems design, the 'human' or cognitive complexity of tasks is not simply a matter of philosophical muse. Enough evidence has been accrued to show that the human predisposition to certain errors, (which may so easily lead to catastrophe) is not trivially related to the 'difficulty' or 'complexity' of a task, as considered with the aid of hindsight (Collins and Leathley 1995).

What do we Actually Mean?

My feeling is that when most of us use the term 'complex' in a technical setting we are not, in fact, using the word in a 'technical' sense. Rather we are using it in the connotative or affective manner of the tautological definitions I mentioned above. We use it to convey a notion of the difficulty of analysis, design, development and maintenance of artefacts (often systems). But I would counsel against this usage; for the simple reason that it is better to call a digging implement a shovel than anything else. I would argue that, if a system is 'difficult to analyse', then it should be referred to as such. If it has many parts that interact, then this should be stated in its description.

Why such pedantry? It is for this reason: one of the views often expressed of truly complex systems is that they exhibit emergent behaviour. Such behaviour may be a system level effect arising from the interaction of many parts. It is problematic that such emergent, system-level behaviours may not admit to the reductionist analytical techniques of the type commonly employed in safety analysis (Collins and Thompson 1997).

A good reason for being restrained in the use of the word 'complex' is that its introduction may (self referentially) make the discussion itself more complex!

Hence the neologism coined in the title of this essay: What is 'Complexologistification'?. Why, simply the process of changing a given subject into a discussion of the study of the science of the complex.

References

Bennett C H (1990). How to Define Complexity in Physics and Why. Complexity, Entropy and the Physics of Information. SFI Studies in the Sciences of Complexity, Vol VIII. W H Zurek (Ed). Addison Wesley: Redwood City, CA.

Casti J L (1994). Complexification: Explaining a Paradoxical World through the Science of Surprise. Abacus : London

Chaitin G (1966). On the Length of Programs for Computing Finite Binary Sequences. Journal of the Association of Computing Machinery. 13:4, 547-569

Collins R J and Leathley B (1995). Psychological Predispositions to Errors in Safety, Reliability and Failure Analysis. Safety and Reliability, 14:3, 6-42 Collins R J and Thompson (1997). Systemic Failure Modes: A Model for Perrow's Normal Accidents in Complex, Safety Critical Systems.   And:
Searching for Systemic Failure Modes. In Advances in Safety and Reliability. Proceedings of the ESREL '97 International Conference on Safety and Reliability, 17-20 June 1997, C Guedes Soares (Ed), Elsevier Science: Oxford

Flood R L and Carson E R (1988). Dealing with Complexity: An Introduction to the Theory and Application of Systems Science. Plenum Press: New York.

Halstead M H (1977) Elements of Software Science Amsterdam: North Holland

Kolmogorov A (1965). Three approaches to the Definition of the Concept 'amount of information'. Problemy Peradachi Informatsii

McCabe T (1976). A Software Complexity Measure. IEEE Transactions on Software Engineering. Volume 2.308-320

Reply to Complexologistification

Kevin Geary

MOD's Procurement Executive, Abbey Wood

This is a reply to the previous article and was published in Safety Systems 7-2, January 1998.

In his article on the meaning of 'complexity' (SCS Club Newsletter, vol 7, issue 1, Sept. 1997, pp 9-10), Rob Collins argues that various definitions were either tautological or indefinitive. I assume Rob's discussion was intended as a challenge to improve upon the complex problem of defining complexity. Well, I will try.

Taking a cue from Rob's proud inclination to ask 'dumb' questions, let us get straight to the point and expose tautological definitions as a strategy for covering up engineering embarrassment. I propose the 'dumb' answer to the 'dumb' question as being that 'complex' means 'nobody can explain how it works'.

Of course, engineers do not build systems they cannot explain! We present diagrams, we analyse, we review, and we do all manner of things that require an explanation of how a system works. But how do we really grapple with a complex system? We apply explanation hierarchically. At a detailed level we explain individual components and interfaces. At subsystem and system levels, explanation adopts a largely topological nature. We can rationalise about any part or all of a complex system, but to do so we have to divide and conquer (i.e., manage) our level of understanding. As an ex-programmer, I recall the concept of a 'head-full' as a measure of maximum module size.

We all pride ourselves on our understanding of and ability to explain the World around us. But how many of us has not been hooked by the inquisitive 5 year old, who's response to everything is 'Why?' When you get beyond the atomic structure of the universe, you resort to an irritated 'Because that's how it

---

© Kevin Geary 1998.
Published by the Safety-Critical Systems Club. All Rights Reserved

is' ... 'Why?'... 'Because it's COMPLEX'. All, that wonderful word for getting out of tight spots.

To qualify my somewhat overly simplistic definition of 'complex', the word is most often used in situations where there is a non-linear relationship between cause and effect. When asked what happens when parameter X is altered by a quantitative amount, the answer may often be, 'It's complex', meaning that the effect on output or behaviour is non-deterministic in that there is no unique predictably quantifiable outcome.

Safety is 'complex'. We can say that because we have (hopefully) engineered out all the linear cause-effect relationships which represent intolerable risk. We duplicate or triplicate safety-critical components or interlocks and, in doing so, introduce 'complexity' into system safety. Of course, what we are really doing is reducing risk by creating a non-deterministic relationship between safety-critical component or interlock failure and the hazard. In the case of duplicated components, there now needs to be a much less likely common mode failure for an accident to be deterministic, hence risk is reduced.

So, next time you are cornered at a conference and asked to define 'complex' you can say, 'Making the system safer by requiring a multiple failure scenario before an accident can occur', and hope that there is no Rob Junior to innocently ask 'Why?'

Demonstrating safety in global navigation systems

Steve Leighton

Siemens Roke Manor Research Ltd

First published in Safety Systems 7-2, January 1998.

The aviation industry is currently faced with an interesting problem revolving around the safety certification of new global navigation systems. The United States government recently offered to the International civil aviation community the free use of its satellite-based navigation system, the Global Positioning System (GPS). GPS allows aircraft with suitable receivers to determine their horizontal position with an accuracy of the order of 67 metres [1] anywhere in the World. It can be received in all weather conditions, 24 hours a day, 365 days a year. It is available in remote, oceanic and mountainous areas where no ground-based navigation aids exist.

In addition, GPS receiving equipment can be of low cost. In short, it offers many benefits to the users. However, GPS has not yet achieved the wide-scale approval that would allow the airlines to reap all the benefits associated with freedom from the ground infrastructure. So what are the problems in gaining approval? The problems associated with the use of GPS are institutional, legal, safety-related and, lastly, technical. This article will focus primarily on the safety-related problems of GPS.

The heart of the problem lies with the operator of the system, the US Department of Defense. GPS was originally built as a military navigation system and, as a consequence, details of its design and internal operation are not publicly available. The only information that the users are provided with is that contained within the 'GPS Signal Specification', a document that amounts to little more than an interface specification. However, GPS offers many potential benefits to its users in the aviation industry: the airlines. Therefore, the national air traffic service providers from many states have been obliged to consider the use of GPS for navigation in their airspace, but, before they may do this they have a considerable number of bridges to cross.

© Steve Leighton, 1998
Published by the Safety-Critical Systems Club. All Rights Reserved

The issues associated with the widespread use of GPS have few precedents to aid in their solution. Consider a hypothetical scenario of a nuclear power station located somewhere in the UK, capable of providing enough high-quality power to support the entire country. It suffers from few, if any, outages and does not charge the users anything. The only drawback is that the operator refuses to release any design information or details of maintenance procedures, or even to ensure that all steps are being taken to prevent accidental release of radiation. All that is available is a public document stating that the power coming out of the plant will be at a constant specified voltage and current. There would be a huge public demand to allow use of the free, high-quality power. But how would the appropriate regulatory and environmental authorities be convinced that the use of this plant was safe when no information was available? Particularly after incidents such as Chernobyl. This is the problem facing the air traffic service providers.

With conventional navigation infrastructure, proof of safety is a relatively straightforward, logical, if sometimes technically challenging, task. Equipment is designed using an appropriate safety management methodology commensurate with the criticality of the application for which it was destined. Appropriate techniques are used during the design process, and supporting analysis is conducted to prove the equipment safe.

For example, within the UK, the latest generation of aircraft precision landing aids, the Microwave Landing System (MLS), was designed with reference to a safety methodology appropriate to the most critical of all navigation aids. Due to the extremely demanding nature of the requirements it was decided that all safety-critical elements of the design would be embodied in hardware, and no software (including embedded code) was allowed. This was primarily due to the difficulties of actually demonstrating that software achieved the overall integrity requirements. Additionally, the MLS monitor systems are themselves monitored to ensure that the probability of the equipment transmitting Hazardously Misleading Information (HMI) is minimised. (In the aviation domain HMI from a landing aid is one of the most dangerous scenarios conceivable.) Such requirements arose from the safety analysis effort that was on-going in parallel with the design process. In particular, detailed FMECA, Fault Tree Analysis, etc., were required to take place. These analyses were then used to form much of the reference material for the System Safety Case that is presented to the regulator for approval. With GPS, limited design information is available upon which safety analysis could be based.

Existing navigation aids only support a limited number of aircraft, and this provides some protection in the event of a failure as only a small number of aircraft will be affected. To make matters worse, with GPS all aircraft would be navigating with one system. Despite there being a number of different GPS satellites, they are all controlled from one central location, creating the possibility for a common mode failure. Hence, to date, demonstrating that GPS is safe has been a task limited to analysis based on the past observed performance of the system. Obviously this is not an adequate solution for such a potentially safety-critical system.

The deficiencies with the safety analysis for GPS are reflected in the limited approval that GPS has received for its use on commercial aircraft. To date, Supplemental Means approval of GPS exists world-wide for en-route navigation, the phase of flight between the end of the initial climb and the start of the final descent. Supplemental Means use of GPS allows the flight crew to use GPS for navigation but alternate systems must be available against which the GPS position output should be compared at all times.

Limited Primary Means certification exists for the use of GPS in oceanic and remote areas. Primary Means certification allows the use of GPS as the main navigation aid, however an alternate system must be provided so that in the event of a loss of GPS position the flight crew have another system to utilise. Obviously in remote or oceanic areas there are few, if any, terrestrial navigation aids, so in such areas GPS could help to improve the safety of the flight.

Traditionally, airways have, where possible, been constructed so that they form routes between ground-based radio beacons. Hence, there is always a navigation aid directly along the aircraft's desired flight path that it may navigate towards. Obviously this approach is restricting. Navigation aids cannot always be located exactly in the optimum locations. In addition some such route structures have become bottle necks during busy periods of the day. It has become necessary to create new routes not defined by direct connections between navigation aids to increase the capacity of the European airspace. The term for flight in such an environment is area navigation. Europe has recently approved the use of GPS for area navigation. GPS is useful for area navigation as it is not dependent upon ground-based aids. GPS receivers can be programmed to provide the flight crew with appropriate guidance to reach any desired location. However, the European GPS area navigation approval does not permit the aircraft operators to withdraw any alternate navigation systems from their aircraft, so there is always another system to revert to should GPS fail.

To overcome the problems associated with lack of available GPS design information, the European air traffic service providers are participating in an ambitious European programme to build what amounts to a monitor for the GPS. This programme is known as EGNOS, the European Geostationary Navigation Satellite System.

EGNOS, as its name implies, provides an augmentation to the basic GPS service that improves the integrity of GPS. It consists of a number of monitor stations located throughout Europe that receive GPS signals. The data they receive is relayed via a communication network to a Mission Control Centre (MCC) which processes the received data. If the MCC detects any failure or anomaly in the performance of GPS an alarm is generated. This alarm is relayed to the aircraft via a geostationary satellite communication link providing Europe-wide coverage. The ambitious goals of the EGNOS project include being able to inform the user of a problem with GPS within 6 seconds of the onset of the failure or anomalous condition. This time includes reception of the GPS signal at the monitor station, transmission and receipt of the data at the MCC (which may be on the opposite side of Europe), determination of the alarm condition, up-link of the alarm to a geostationary satellite, and down-link and processing by the user equipment.

Many European ATC service providers are involved in the EGNOS programme to ensure that the system is designed in a manner consistent with the safety-criticality of the aviation operations it must support. This time the design information is available and the necessary safety analysis is currently ongoing.

EGNOS is possibly the most complex monitor system yet developed and encompasses a variety of disciplines from space system engineers, ATC system engineers, communication specialists, and navigation specialists, to name but a few. All are working together to ensure that a system is designed in which safety is the paramount concern. With the implementation of EGNOS it will be possible for the first time to have a primary means global satellite-based navigation system that will support aircraft navigation both en-route and also for some forms of precision approach to airfields. Similar activities are also on-going in the USA and Japan.

The issue of the safety certification of satellite navigation systems is a very high profile topic within the ATC infrastructure provision and regulation domain. A great deal of effort is being expended by all parties to ensure that when EGNOS becomes operational, safety analysis has been undertaken to demonstrate that it is safe not only to the UK aviation regulator but also to the regulatory bodies of every other European state which will have the benefit of the EGNOS service. For the first time ever we will have taken a global navigation system that we could only strictly consider to be unsafe (having been unable to undertake adequate safety analyses) and we will have made it safe by the addition of a complex Europe-wide monitoring system. An achievement indeed, all driven, ultimately, by the need for greater air traffic control capacity, cost savings and safety.

Reference

[1] 'Assessing the Navigation Data Input to Flight Management Systems', P Nisner, Royal Institute of Navigation, November 1997.

Software as goods:
some answers, but yet more questions

Ian Lloyd

The Law School, University of Strathclyde

First published in Safety Systems 6-2, January 1997.

Introduction

The community charge, more commonly known as the poll tax, proved one of the less popular forms of taxation in recent British history. In fiscal terms, the tax is no longer operative, but legal issues concerned with its operation continue to come before the courts. The recent case of St Albans v. ICL is likely to prove a landmark decision in the field of software liability and sheds new light on the vexed question of whether software might be regarded as a product. The result of the case is to impose a high degree of liability upon software producers and also to cast doubt on the effectiveness of the exclusion and limitation software contracts.

Facts of the Case

The key (and controversial) element of the poll tax was that, subject to a very limited number of exceptions, all those aged 18 and above living in a local government district were required to pay a single sum by way of poll tax. In administrative terms, this approach simplified the task of the local authorities. Effectively, all that was required was to calculate the income required, the number of persons liable to pay the tax and divide the one by the other.

If ever a task could be seen as made for the computer, this was surely it, and without exception local authorities invested heavily in IT systems to administer the tax. Many of the authorities, St Albans included, entered into contracts with the computer supplier ICL who promoted an IT system referred to as The ICL Solution'. At the time the contract was signed, the elements of the system re-

© Ian Lloyd, 1997
Published by the Safety-Critical Systems Club. All Rights Reserved

quired to cope with the specific demands of the community charge had not been completed or tested. This fact was promoted as a positive benefit to the authority. The developers would use a 70 strong development team to produce the necessary software and by entering into the contract the Council would be able 'to input into the development process in order to be sure that this product meets your specific requirements'.

The contract, valued at some £1.3 million, was concluded subject to ICUs standard terms and conditions, which limited liability for losses to a maximum of £100,000. The system was delivered to the Council timeously but, as envisaged in the contract, the software required was to be supplied and installed in stages as various elements were completed, and in line with legislative requirements relating to the introduction of the new tax. Initial elements were to be completed in Autumn 1988 with the full system being operable by February 1990.

One of the first tasks which required to be conducted by local authorities was to calculate the numbers of persons in their area liable to pay the tax. The calculation was carried out using the ICL system in early December 1989, and a figure of 97,384 was produced. Unfortunately, the version of the software used had a bug and for some unknown reason a new release which would have cured the problem was not installed on the Council's computers prior to the calculation. The correct figure, it was subsequently discovered, was almost 3,000 lower at 94,418. The financial effects were significant. The Council were effectively caught in a double-edged trap. Their income was reduced because the 3,000 phantom taxpayers would clearly not produce any income. To compound matters, part of the community charge income was destined to be transferred to the larger Hertfordshire County Council and this figure was also calculated on the basis that St Alban's taxpaying population was greater than it actually was. When the accounts were finally completed, it was calculated that the loss to St Albans was over £1.3 million. Legal proceedings were brought against ICL seeking recovery of this sum.

Legal Issues

Two key issues were raised before the courts. First, was the defendant in breach of their contractual obligation? Second, were its exclusion and limitation clauses effective? The High Court found in favour of the plaintiff on all counts and an appeal was made to the Court of Appeal.

Although ICL did not dispute the fact that the software involved in the calculation had been defective, they argued that their obligation was merely to supply a system which would be fully operative at the end of February 1990. Until then, as was recognised in the contract, the system would be in the course of development. Save where it could be shown that the supplier had acted negligently, it was argued, *'the plaintiffs had impliedly agreed to accept the software supplied, bugs and all'*. This contention was rejected by the court which held that:

> *'Parties who respectively agree to supply and acquire a system recognising that it is still in the course of development cannot be taken, merely by virtue of that recognition, to intend that the supplier shall be at liberty to supply software which cannot perform the function expected of it at the stage of development at which it is supplied.'*

This is a significant finding. The fact that a system is - to the knowledge of both developer and customer - under development, will not provide an excuse for the developer if its performance falls below a satisfactory standard. The Court of Appeal went further. In his judgment, Sir Iain Glidewell posed the question 'Is software goods?' He answered:

> *'If a disc carrying the program is transferred, by way of sale or hire, and the program is in some way defective, so that it will not instruct or enable the computer to achieve the intended purpose, is this a defect in the disc? Put more precisely, would the seller or hirer of the disc be in breach of the terms of quality or fitness implied by s.14 of the Sale of Goods Act?*
>
> *'Suppose I buy an instruction manual on the maintenance and repair of a particular make of car. The instructions are wrong in an important respect. Anybody who follows them is likely to cause serious damage to the engine of his car. In my view the instructions are an integral part of the manual. The manual including the instructions, whether in a book or a video cassette, would in my opinion be 'goods' within the meaning of the Sale of Goods Act and the defective instructions would result in breach of the implied terms. 'If this is correct, I can see no logical reason why it should not also be correct in relation to a computer disc onto which a program designed and intended to instruct or enable a computer to achieve particular functions has been encoded. If the disc is sold or hired by the computer manufacturer, but the program is defective, in my opinion there would prima facie be a breach of the terms as to quality and fitness for purpose implied by the Sale of Goods Act.'*

The same result, he suggested, would apply where software is not supplied on a disc or tape. It might, for example, be downloaded over the Internet so that no disc or tangible object is transferred to the acquirer.

The effect of this dictum would be to impose a high degree of liability upon software producers and suppliers. Faced with such extensive liability, the attractions for suppliers of exclusion or limitation clauses are readily apparent. Here again, the court's findings offer scant consolation. In the United Kingdom, exclusion and limitation clauses are either prohibited or subjected to the application of a test of reasonableness in all cases involving consumers. For non-consumer contracts, exclusion clauses will be upheld except where the contract is classed as a standard form contract. In the St Albans' case, the Council published a call for tenders, negotiated — albeit fairly incompetently — with a number of potential suppliers, engaged in further negotiations with ICL and concluded a contract, one clause of which stated that it was subject to ICL's standard terms and conditions. This contract was held to be a standard form contract and, upholding the findings of the trial judge, the Court of Appeal held that it did not satisfy the statutory criterion of reasonableness.

Where next?

Until St Albans v. ICL, courts have gone to great lengths to avoid giving a direct answer to the question whether software should be considered as goods or services. As indicated in the case, the prime consequence of holding software to be goods will be that a software developer will not be able to argue that all software has defects or that best efforts were applied in order to develop a satisfactory system. The court will look only at what has been promised and at what has been delivered. Prudent developers will take care to make sure that they do not promise more than they can deliver, and wise customers will ensure that their expectations are specified clearly. The requirements that software be of satisfactory quality and fit for the purpose for which it is supplied do not require perfection and further cases will be required to give a clearer understanding of what these concepts mean in a software context. The case does undoubtedly, however, send a cautionary signal to software developers.

Safety and the Year 2000

Tony Foord

    4-sight Consulting

First published in Safety Systems 6-3, May 1997.

All hype or a real problem? You could be forgiven for thinking it was A.D. 999 if you read today's computing press. Many in A.D. 999 were convinced that the world would end on 1 January 1000. A thousand years later the prophets of doom are out in force again. I am told, for example, that the PC on which I am typing this article will change into a 4 January 1980 pumpkin at the stroke of midnight on 31 December 1999. What I am not told is that when I start work and power up my PC on 2 January 2000, a single instruction to reset the date will fix the problem. In fact for most people the problem will be fixed automatically for them by the server that is connected to their PC across the LAN.

I am also not told that this same PC will quite correctly move from 28 to 29 February to 1 March 2000 because the simple 'divide by 4' algorithm works perfectly for the Year 2000. It will fail on 1 March 2100, but I will leave that for my great grand-children to sort out. So, is it all hype? Surely safety-related systems are secure from Year 2000 or other date-related problems? Unfortunately the answer is no. It's not all hype and time- or date-related problems go back a long way and do include safety-related systems. A combination of official secrets, commercial confidentiality and embarrassment mean that examples are not easy to come by, but they do exist.

Examples of problems with time or dates

An early computer
On 1 January 1972 a KDF9 computer failed to operate when powered up. Eventually they tried resetting the date to 31 December 1971 and all was well. Examination of the code revealed that the date algorithm written in the early 60s could handle only the next 10 years. No-one in the early 60s expected a KDF9

© Tony Foord, 1997
Published by the Safety-Critical Systems Club. All Rights Reserved

to be in use into the next decade. The commercial and safety implications were not significant.

Aviation

The pilot of a 4-engined plane out on patrol was told to shutdown 3 engines and return to base immediately. As a result of confusion over dates, components that were outside their stated working life had been installed in his engines. Dangerous problems had been encountered on other engines where the same components had been allowed to exceed their stated working life. The commercial implications were significant, involving several thousand pounds of refit work. The action was precautionary and there was no harm to people.

Process industry

A computer was controlling a batch reaction on a chemical plant during the night when summer time ended and the clocks had to be put back one hour. The operator reset the clock in the computer so that the time indicated 2 am instead of 3 am. The computer then shut the plant down for an hour until the clock reached 3 am again [1]. The hour of lost production was significant commercially, but there was no harm to people.

Refining

Last year the control system operating a smelter in New Zealand failed on 29 February 1996. It was necessary to shut-down and there was no harm to people, but the cost of the unplanned shut-down was estimated at £1M. The control system could not handle a leap year. Like the KDF9, having the wrong date had wider implications and prevented the whole system from working properly.

Medicine

A lady had an emergency pace-maker fitted while on holiday in Canada in November last year. She recovered very quickly from the operation and the doctors there said she was fit to travel home to England, but that she must have the pace-maker checked and regulated within one month. On consulting her doctor back home 2 weeks later she was rather surprised to be given a hospital appointment for April 1997. When she asked her doctor to check this, he discovered that her Canadian notes stated the operation had been done on 11/4/96. This was interpreted as April (not November) so an annual check-up had been arranged for the next April, rather than the essential one month check-up for December [2]. The commercial and safety implications were not significant, but only because the patient challenged what she was told by her English doctor!

Transport
Many transport systems now use global positioning satellites (GPS) to identify the location of moving vehicles, for example trains. Some parts of the GPS system will not function after 25 August 1999 as the date field will run out of bits (the KDF9 problem again). Both GPS and the safety-related signalling systems that rely on it are receiving urgent attention.

Issues with Dates

Standards
So why are there issues about dates? The problems lie both in the lack of standards and in the way the standards themselves define dates. North Americans use mm/dd/yy and Europeans use dd/mm/yy. The standards for COBOL, barcodes, etc. insist that the year field has only two digits. This is not just a problem for supermarkets. Barcodes are used on components in aviation and in other industries to indicate shelf life. During use, a component's working life may be recorded in monitoring systems which use only two digits for the year.

The standard time transmitted by the National Physical Laboratory (NPL) from Rugby uses years with only two digits. A recent newsletter [3] notified users of the NPL transmission about the next leap second and also warned that it was up to users to ensure that their equipment would handle a year of '00'.

Leap years are not well understood. Even though my PC will handle 29 February 2000, at least one mainframe operating system currently in use will not. The systems programmer mistakenly thought that the centuries were never leap years. Fixing this mistake is currently costing several million pounds.

Weak typing and short cuts in coding
Many commercial IT systems contain legacy code written long before the need for strong typing was understood. Some legacy systems include 'short cuts' in coding, such that '99' or '00' in the year field is used to terminate a data stream. In other systems '19' or '199' is hard-coded as a prefix to all years. I have not yet found occurrences of this in safety-related systems, and I would welcome sources of real-world examples.

Demand rate from control systems
We should not restrict our concern to safety-related systems. The design of protection systems is based on assumptions about demand rates. If the underlying control systems are not 'year 2000 compliant' then problems with dates in the control systems could significantly affect the demand rate on the protection systems. For example, many SCADA systems are PC based. Those PC systems

that operate continuously may not be corrected as easily as suggested in the first section above.

Common-cause failures

Resilience is often enhanced by redundancy and diverse back-up systems. Nearly all systems use dates somewhere, so all the diverse and redundant systems will need to be Year 2000 Compliant or else there will be the potential for a significant common cause of failures.

Configuration control

As with all safety-related systems, it is not just the application code which matters. The code libraries, operating systems, tools and other parts of the environment have an impact on the performance of the system. Configuration control tools almost always use dates, and these tools must also be Year 2000 Compliant.

Ambiguity, arbitrary limits and exception handling

As the examples show, issues with dates have already arisen and could arise again in safety-related systems. In summary, the issues are the all-too-familiar problems of ambiguity and arbitrary limits. Unfortunately the impact of errors in dates is not restricted simply to printing or recording the wrong day. The errors may have a direct impact because a wrong duration or order is calculated (e.g. the 99 version of the software may be used because it appears more recent than the 00 version). Alternatively, the errors may have an indirect effect because they raise an exception (such as a negative duration) which is not well handled (as for the KDF9, the batch process, the smelter and GPS).

Definition of Year 2000 Compliance

BSI has produced a Code of Practise [4] and a clear definition of Year 2000 Compliance [5] of which a summary is:

Year 2000 conformance shall mean that neither performance nor functionality is affected by dates prior to, during and after the year 2000. In particular:

Rule 1: No value for current date will cause any interruption in operation;
Rule 2: Date-based functionality must behave consistently for dates prior to, during and after year 2000;
Rule 3: In all interfaces and data storage, the century in any date must be specified either explicitly or by unambiguous algorithms or inferencing rules;
Rule 4: Year 2000 must be recognised as a leap year.

For many systems, a range of dates of 99 years is sufficient. An inferencing rule can then be used: e.g. a year XY less than 50 means year 20XY, while a year YZ greater than 50 means 19YZ.

Conclusions

There is not space here to cover the many complex audit, commercial, contractual, insurance and legal issues which need to be considered as part of the solution. As with any change, there is always the risk that the solution will introduce new problems. An essential starting point is the register of safety-related systems — which should already be available within any safety management system. An action plan should include at least the following:

- Inventory identifying all those safety-related systems that include software or firmware (an inventory should include the support software for these systems);
- Review to decide whether it would be cheaper to replace rather than audit;
- Assessment of tools required for audit, analysis and correction;
- Audit to identify where dates are used;
- Impact analysis to show how these dates affect the systems;
- Estimates of time needed for, and cost of, further analysis and correction;
- A second review to reconsider replacement now that costs are known more accurately;
- Interfaces and bridging to integrate corrected software;
- Testing and trials of corrected software.

Working with suppliers and maintainers of the systems is essential. There are questionnaires available to assist in identifying systems that may have problems with dates. There are various commercial tools for checking code to identify where dates are used, and for doing impact analysis. More details of these can be found in [6] and from Web pages, e.g. [7-10].

A key issue is testing for Year 2000 Compliance. The 'Engineering Workstation' or 'Test and Development' system associated with each safety-related system should allow an exact simulation of the operational system. BSI [5] suggest that the tests should cover the feasible lifespan of the system, the full range of dates that need to be stored, and certain critical dates (e.g. leap years and end-of-year). For safety-related systems, the tests should include the exception handling in cases of human or other errors.

Substantial effort and a realistic budget is needed to tackle these issues. For IT systems the cost of achieving Year 2000 Compliance is quoted as about $1 per line of code, and about 50% of that cost is in testing. Co-operation within the SCSC to identify the real problems and eliminate the hype could substantially reduce the cost to industry of tackling these issues.

References
1. Trevor Kletz, 'Computer Control and Human Error', IChemE, 1995, page 20
2. Stuart Fieldhouse, 'Year 2000: Still crazy after all these years', Computer Personnel, February 1997, page 54
3. NPL Newsletter, January 1997
4. DISC PD2000-2, 'A Code of Practice for Year 2000 Management', BSI
5. DISC PD2000-1, 'A Definition of Year 2000 Conformity Requirements', BSI
6. 'The Year 2000, a practical guide for professionals and business managers', BCS, ISBN 0 901865 7 4
7. www.weblaw.co.uk/yr2000.htm
8. www.open.gov.uk/ccta/mill/mbhome.htm
9. www.cssa.co.uk/cssa/new/millen.htm
10. www.bcs.org.uk/millen.htm

Acknowledgement and Disclaimer
While I am very grateful for examples provided by colleagues, including David Spinks of AEA Technology and Prof. Trevor Keltz, the views expressed in this article are entirely personal.

The millennium timebomb (the Y2K problem) — a consultant's dream, or a real problem?

Ray Ward

Health and Safety Executive

First published in Safety Systems 7-2, January 1998.

Are we approaching 'Armageddon' for our computer-based society when we roll over into the new millennium on 31st December 1999, as some consultants would have us believe, or not? Are the pundits scare-mongering for business, or do we have a serious problem with computer systems which industry is not yet properly addressing?

The real answer is that there is a problem but we do not yet know how serious it will be. The Health and Safety Executive (HSE) is trying to find out and is proposing a strategy which will help British industry. The extent and possible effects of the problem for safety-related control systems can only be established by thorough investigation, and a reasoned approach is essential for its solution.

The problem goes back to early computer systems where memory storage was at a premium. Programmers were encouraged to save space and eliminate redundant data wherever possible. Date information was therefore stored with a 'two-digit year'. A 'DDMMYY' date format saved 25% of storage over 'DDMMYYYY' when the digits '19' were assumed. In this convention, 1965 was represented as 65, 1993 as 93; 2000 will be 00, but 1900 is also 00. Computer systems which do recognise 00 as a valid date may respond as if 1900 is meant. Although, in the late 1990s, storage costs are now much lower, vast amounts of stored data and program instructions are still in the old format.

Few people commissioning or writing computer code in the 1970s and 1980s thought that their systems would survive into the 21st century. However, many organisations are still using systems which have been in operation for 15 years or more, and some may be as old as 25 years. These are the so-called 'legacy' systems.

The change from 1999 to 2000 is the most readily recognised problem, but you should be aware that there is a more general problem associated with what

© Ray Ward, 1998
Published by the Safety-Critical Systems Club. All Rights Reserved

is called 'date discontinuity'. Date discontinuity occurs when the time (as expressed by a system or its software), does not successfully move forward in line with true time.

For instance, some software systems are equipped with 'clocks' which calculate time from a fixed point (e.g. by counting the number of clock ticks since 1st January 1980 or some other arbitrary date of significance to the manufacturer). When the register which accumulates these clock ticks is full, it will overflow (like a car odometer) and show zero. This will be interpreted by the software as the fixed date of origin, i.e.1980.

Other date-related problems may also arise. For example:

- some systems use '99' in the year field to indicate 'end of run', which may give a problem at 9/9/99;
- some use '00' in the year field to indicate 'invalid record';
- some systems will not 'know' that the Year 2000 is a leap year, because century years are not usually leap years and the rules governing nomination of leap years are not common knowledge. Thus the roll-over from 29th February to 1st March 2000 and/or moving to the 366th day (31st December 2000) may cause a problem.

Implications

The implications for safety are far reaching in that many basic functions of computer systems will fail. All calculations and processes based on two-digit year data are likely either to fail completely or to return incorrect results. This could be in sorting routines, validation routines or in any automatically generated date-based information (time stamps, receipt dates, automatic indexing, reference numbers, etc.) or at the interfaces between systems. Examples include:

- lists of outstanding actions;
- diagnostic data;
- database calculations;
- changes to security information where complete failure and/or the transfer of ambiguous data may occur.

If this range of examples were not enough, a big problem for users will be in trying to establish whether or not they even have a Y2K problem! Finding out is estimated by some experts as 40% of the total project cost.

The Law

Suppliers, employers, the self-employed and consultants have legal obligations. These are set out in general terms in the Health and Safety at Work etc ... Act 1974 (HSWA) and the Supply of Machinery (Safety) Regulations 1992 as amended. Other aspects relating to consumer safety are covered by the Consumer Protection Act 1987 (CPA), but HSE's concern is safety in the workplace.

Duty of employers:
Section 2 (1) of HSWA in particular places a duty on employers to ensure, so far as is reasonably practicable, the health, safety and welfare of their employees. Section 3 of HSWA requires employers and the self-employed to ensure, so far as is reasonably practicable, the health and safety of others who may be affected by their work activities. This also covers the work of consultants who provide advice and technical support to their customers.

Other, more specific legislation, relating to control systems is the Provision and Use of Work Equipment Regulations 1992 (PUWER). Regulation 5 requires that work equipment must be suitable, by design, construction or adaptation, for its intended purpose. Regulation 18 requires that employers ensure, so far as is reasonably practicable, that the operation of a control system does not pose any additional risk to health or safety. Any hazardous event which may result from an uncorrected date discontinuity problem in a safety-related control system may therefore contravene this regulation.

Duty of designers, manufacturers and suppliers:
Those who design, manufacture or supply articles for use at work also have a duty to those who use those articles (section 6 of HSWA). This duty extends to the information provided for use. It also extends to the revision of such information if it becomes known that anything gives rise to a risk to health or safety. In other words, the law requires designers, manufacturers and suppliers to be proactive and to take reasonably practicable steps to inform their customers of potential problems once they become known. Date discontinuity problems in the hardware or software of a safety-related control system is such a problem.
Manufacturers and suppliers also have more specific duties under the Supply of Machinery (Safety) Regulations 1992 as amended, for machines that have control systems. In particular, a fault in the control circuit logic, such as a date discontinuity problem, should not lead to dangerous conditions.

What HSE is doing

1. A research contract has been let to Real Time Engineering Ltd., which will concentrate on manufacturing and process-control systems where safety is dependent upon the computer. The report from this research is to be published soon.

2. The research report will be followed in March 1998 by a free HSE guidance leaflet directed primarily at SMEs.
3. The Commons Select Committee on Science and Technology is currently investigating the issues surrounding the Year 2000 problem. Along with many other organisations, HSE has been asked to submit evidence to this Committee with respect to the Y2K problem as it affects safety-related control systems.
4. HSE will investigate any incidence of date discontinuity which is brought to its attention so that we may add to the fund of knowledge of how to deal with this problem.

Reverse Engineering the Software Design of a Safety-related ATM System

Ron Pierce and Mary Johnston

CSE International Ltd

First published in Safety Systems 22-2, January 2013.

1 Introduction

ARTAS - the ATM (air traffic management) suRveillance Tracker And Server - is an advanced multi-surveillance data-processing system which accepts plot and target report information from different surveillance sensor technologies and combines it into an integrated air situation picture. ARTAS accepts primary, secondary and Mode S radar, automatic dependence surveillance and multi-lateration inputs and can serve many user systems - users can select the tracks which they wish to receive on the basis of various filtering criteria. For readers unfamiliar with modern surveillance technology, there are some brief explanatory notes at the end of this article.

ARTAS was developed over a number of years by EUROCONTROL to promote ATM harmonisation and integration on the continent and is being maintained by the EUROCONTROL central maintenance and support organisation, with the help of an industrial partner, currently COMSOFT GmbH. ARTAS is used operationally by a large number of air traffic control service providers and further organisations are planning to put it into service. If you fly over Europe, there is a good chance that ARTAS will at some stage be providing the surveillance data to the air traffic controller responsible for the safety of your flight.

In 2005 CSE International Ltd undertook a safety assessment of ARTAS. While this assessment was generally positive, it identified one particular deficiency in the safety evidence available, specifically the fact that the legacy software had been developed in the 1990s to military quality standards and not

© Ron Pierce and Mary Johnston, 2013
Published by the Safety-Critical Systems Club. All Rights Reserved

to a recognised software safety standard. The integrity target for the software is Software Assurance Level 3 (SWAL 3) in accordance with the EUROCONTROL Recommendations for Air Navigation Services (ANS) Software [1].

The EUROCONTROL Surveillance Products and Services Unit subsequently undertook a programme to provide the missing software assurance evidence. The system and software requirements have been re-structured to bring them into line with best practice requirements engineering, and an improved set of tests, traceable to the requirements, has been developed. New system components (for example, the new man-machine interface and a new middleware) have been developed to SWAL 3 from the start.

A further key requirement to meet SWAL 3 is a documented Software Architectural Design (SDD) with bi-directional traceability to the software requirements. Under a contract from EUROCONTROL, CSE and COMSOFT have produced new SDDs for ARTAS. The six ARTAS Computer Software Configuration Items (CSCIs) for which SDDs were required are written in Ada 83. They are mostly large, contain hundreds of Ada compilation units, and have a complex internal structure, all of which features made reconstructing an architectural design a major exercise.

This article outlines how the SDDs were created and describes some of the benefits and challenges of the approach taken.

2 Method and Notation

A software architectural design has two aspects. The first is the static structure of the code - how it is divided into components, the relationships between those components, and the nature of the interfaces offered by the components. The second aspect is the dynamic behaviour of the software - how the components interact when the software is running.

There are two basic approaches which can be considered when attempting to document the static structure of a software system for which the source code already exists:

- by inspection of the source code (possibly aided by source code browsing tools) and manual construction of a design model; or
- by automatic reverse engineering using a software tool to capture information from the source code and automatically populate a design model.

In the case of the dynamic behaviour, automatic reverse engineering would rarely, if ever, be possible and the dynamic descriptions had to be created and maintained manually.

To facilitate maintainability of the SDD documents and the consistency of the static structural description with respect to the source code, the decision was taken to automatically reverse engineer the static part of the architectural design. One merit of this approach is that all the entities that are candidates for traceability links from the requirements are certain to be included in the model. Unified Modelling Language (UML) [2] was chosen as the notation in which to express the design, supported by IBM's Rhapsody tool, which has an Ada to UML reverse engineering capability. Other design methods were considered, but the selection was made on the grounds that UML is widely known and has a rich set of notations for expressing both static structure and dynamic behaviour. The Rhapsody tool was chosen partly for its reverse engineering capability and partly from EUROCONTROL's expressed preferences.

3 Describing the Design

Major software components of the design (computer software components or CSCs) are mapped to UML packages. Ada packages are mapped to UML classes with the visible subprograms becoming the operations of the class and the visible data declarations the attributes. The stereotype< > is applied to the majority of these classes showing that there is only one instance at run-time. Ada generic packages are mapped to UML template classes.

The dependencies between the UML classes are represented by the stereotype <<usage>> (representing the with clauses of the Ada compilation units), and generic instantiations by the stereotype <<binds>>.

Package diagrams and object model diagrams are used to depict the dependencies between UML packages and between classes. An example of a package dependency diagram is given in Figure 1. It was necessary to be selective with the diagrams produced and the information contained in each diagram because of the large number of classes and dependencies in the code.

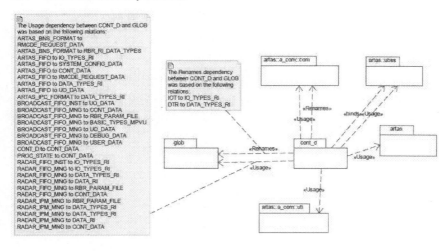

Fig. 1. Example of external package dependencies diagram

In describing the dynamic behaviour of each CSCI, our aim was to explain important, difficult or obscure areas of the design for the benefit of new staff, who may have to maintain the software. Data flow diagrams (object model diagrams with data flows indicated by stereotypes), StateCharts, sequence diagrams and interaction diagrams were all employed to good effect.

SDD documents are automatically generated from the model information, using a combination of standard tool facilities and Word macros; a relatively small amount of manual intervention is needed to create a finished SDD document.

4 Traceability

The IBM DOORS® tool, which has an interface to Rhapsody, is used for requirements traceability. Traceability was sought between each software requirement and the corresponding entity or entities in the SDD model which implement the requirement. This had to be done by inspection of the source code, and a good knowledge of the code was helpful here. Generally a functional requirement is linked to specific operations of a class, or in some cases to the complete class. Where a requirement could not be traced to a class or operation (for example, because it was a non-functional requirement concerning performance or capacity), it was traced to a Rhapsody comment explaining why traceability could not be achieved. Generating forward and reverse traceability matrices was accomplished by the standard facilities of DOORS and Rhapsody.

5 Gap Analysis

Once the traceability matrices were created, a gap analysis was undertaken to determine any discrepancies between the software requirements and the implementation (for example requirements not implemented, or functionality not warranted by a requirement). The gap analysis was performed using the forward and reverse traceability matrices and also by examination of the dynamic behaviour as expressed in the UML design to ensure that it was consistent with the Software Requirements Specification (SRS).

In general, very few gaps were discovered, which is reassuring in a safety-related system which is so widely used operationally. The major gap concerned error and exception-handling behaviour, which is not specified in the SRSs, but is of importance in a safety-related system. A set of additional SRS requirements to fill this gap was proposed. Other requirements and system design documentation changes, to address minor gaps, were also proposed. However, the gap analysis did not reveal any unwanted functionality that could be detrimental to the safety of the ARTAS system.

Some apparently unused code was discovered and this will need to be studied to decide whether it should be removed or left in place, possibly protected by diagnostic outputs, which would alert maintainers if any of the suspected "dead" code were to be executed.

6 Verification and Safety Audit

Although the creation of the SDDs and the insertion of the traceability information is only part of the total software lifecycle, and was performed retrospectively, it was still necessary to perform the relevant verification and safety audit activities. Verification of the SDD and traceability matrices was performed by an independent person. In particular, the completeness and correctness of the traceability information were checked by the verifier and the traceability links corrected where errors were found. A further check was to take samples of the overall system level (System- Segment Specification or SSS) requirements and trace these to the software requirements specification (SRS) and thence into the software architecture represented by the SDDs.

An independent safety audit was also performed to confirm that all the provisions of the ANS Software Recommendations [1] had been met. Finally, a Software Safety Folder containing the SDDs, review reports, verification reports and safety audit report was created to capture the output of the project in accordance with the requirements of the ANS Software Recommendations.

7 Conclusions

It has been possible to use the reverse engineering capabilities of a commercially available software engineering tool to create a software architectural design expressed in UML for a set of very complex Ada 83 programs, using a combination of automated methods and manual construction. As well as satisfying the requirements of the relevant software safety standard and allowing traceability to the design and code to be realised, the resulting SDDs provide a useful addition to the system design documentation, which will be helpful to anyone trying to increase their understanding of the software.

Notes on Surveillance Technologies

A primary radar senses the presence of an aircraft by reflecting electromagnetic waves (usually in the microwave region) from the aircraft's skin. A secondary radar uses a transponder on the aircraft to reply to an interrogation pulse from the ground station (and will send the aircraft identity and height information in the reply). A Mode S radar is a development of the secondary radar and a Mode S unit selectively interrogates each aircraft within its coverage in turn; this greatly reduces the problem of garbled signals, which can occur with conventional secondary radar and allows a great deal more information about the aircraft's state and intentions to be included in the reply. Automatic dependent surveillance (ADS) is an alternative system in which the aircraft sends its own position to one or more ground stations. Multilateration is a system in which the position of an aircraft is measured by the differences in arrival time of radio signals from the aircraft at three or more ground stations. Multilateration and ADS do not require the rotating antennas used by primary, secondary and Mode S radars.

Acknowledgements The authors would like to thank their colleagues at CSE and COMSOFT who took part in this project, the COMSOFT project manager Robert Clauss, Markus Schöpflin, Uwe Langer, Guillaume Dunoyer, Derek Summers and Jenny Butler. The ARTAS team at EUROCONTROL also made an important contribution to the success of this project.

References
[1] EUROCONTROL, Recommendations for ANS Software, SAF.ET1.ST03.1000. GUI-01-00 Edition 1.0, 2005
[2] Object Modelling Group (OMG). Unified Modelling Language (OMG UML), Infrastructure, Version 2.2 and Unified Modelling Language Superstructure, Version 2.2. OMG reference: formal/2009-02-04

AUTHOR INDEX

Rob Collins	127
Philip Cosgriff	95
Tony Foord	141
Ken Frith	1
Kevin Geary	131
Ian Gilchrist	65
Mary Johnston	151
Tim Kelly	5, 99
Steve Leighton	133
Nancy G. Leveson	19
Ian Lloyd	137
Guy Mason	83
Chris Moore	25
Mark Nicholson	117
Odd Nordland	89
Ron Pierce	151
Stan Price	15
Andrew Rae	117
Felix Redmill	31, 39, 53
John Ridgway	9, 59
Brian Sherwood-Jones	79
David J Smith	73
David Ward	105, 111
Ray Ward	147

Many thanks to Brian Jepson for the scanning of old newsletters.